Tribology for Engineers

Related titles:

Tribology and dynamics of engine and powertrain: fundamentals, applications and future trends
(ISBN 978-1-84569-361-9)
Tribology, the science of friction, wear and lubrication, is one of the cornerstones of engineering's quest for efficiency and conservation of resources. *Tribology and dynamics of engine and powertrain: fundamentals, applications and future trends* provides an authoritative and comprehensive overview of the disciplines of dynamics and tribology using a multi-physics and multi-scale approach to improve automotive engine and powertrain technology. Part I reviews the fundamental aspects of the physics of motion, particularly the multi-body approach to multi-physics, multi-scale problem solving in tribology. Fundamental issues in tribology are then described in detail, from surface phenomena in thin-film tribology, to impact dynamics, fluid film and elastohydrodynamic lubrication means of measurement and evaluation. These chapters provide an understanding of the theoretical foundation for Part II which includes many aspects of the physics of motion at a multitude of interaction scales from large displacement dynamics to noise and vibration tribology, all of which affect engines and powertrains. Many chapters are contributed by well-established practitioners disseminating their valuable knowledge and expertise on specific engine and powertrain sub-systems. These include overviews of engine and powertrain issues, engine bearings, piston systems, valve trains, transmission and many aspects of drivetrain systems. The final part of the book considers the emerging areas of microengines and gears as well as nano-scale surface engineering.

Tribology of natural fiber polymer composites
(ISBN 978-1-84569-393-0)
Tribology of natural fiber polymer composites examines the availability and processing of natural fiber composites and their structural, thermal, mechanical and tribological properties. It explores sources of natural fibers, their extraction and surface modification as well as properties of chemically modified natural fibers. It provides an overview of the tribology of polymer composites and the role of fiber reinforcement and filters in modifying tribological composites.

Solving tribology problems in rotating machines
(ISBN 978-1-84569-110-3)
Solving tribology problems in rotating machines is an essential reference for engineers involved in the design and operation of rotating machines in such sectors as power generation, electrical and automotive engineering. Bearings are widely used in rotating machines. Understanding the factors affecting their reliability and service life is essential in ensuring good machine design and performance. *Solving tribology problems in rotating machines* reviews these factors and their implications for improved machine performance. The first two chapters review ways of assessing the performance and reliability of rolling-element bearings. The author then goes on to discuss key performance problems and the factors affecting bearing reliability. There are chapters on cage and roller slip, and particular types of failure in equipment such as alternators, condensers and pumps. The author also reviews the effects of such factors as localised electrical currents, seating, clearance, grades of lubricant, axial forces, vibration on performance and service life. The book concludes by reviewing ways of improving bearing design.

Details of these and other Woodhead Publishing books can be obtained by:

- visiting our web site at www.woodheadpublishing.com
- contacting Customer Services (e-mail: sales@woodheadpublishing.com; fax: +44 (0) 1223 832819; tel.: +44 (0) 1223 499140 ext. 130; address: Woodhead Publishing Limited, 80 High Street, Sawston, Cambridge CB22 3HJ, UK)

If you would like to receive information on forthcoming titles, please send your address details to Customer Services, at the address above. Please confirm which subject areas you are interested in.

Tribology for Engineers

A practical guide

EDITED BY
J. PAULO DAVIM

WP
WOODHEAD
PUBLISHING

Oxford Cambridge Philadelphia New Delhi

Published by Woodhead Publishing Limited, 80 High Street, Sawston,
Cambridge CB22 3HJ, UK
www.woodheadpublishing.com

Woodhead Publishing, 1518 Walnut Street, Suite 1100, Philadelphia, PA 19102-3406, USA

Woodhead Publishing India Private Limited, G-2, Vardaan House, 7/28 Ansari Road,
Daryaganj, New Delhi – 110002, India
www.woodheadpublishingindia.com

First published 2011, Woodhead Publishing Limited

British Library Cataloguing in Publication Data
A catalogue record for this book is available from the British Library.

Woodhead Publishing ISBN 978-0-08-101491-2

The publisher's policy is to use permanent paper from mills that operate a sustainable forestry policy, and which has been manufactured from pulp which is processed using acid-free and elemental chlorine-free practices. Furthermore, the publisher ensures that the text paper and cover board used have met acceptable environmental accreditation standards.

Typeset by RefineCatch Limited, Bungay, Suffolk

Contents

Preface

The term tribology derives from the Greek '*tribein*' meaning 'to rub', and '*logos*' meaning 'principle or logic'. Tribology is the 'science and technology of interacting surfaces in relative motion and of associated subjects and practices'. It includes the research and application of principles of friction, wear and lubrication. Nowadays, tribology on the small scale and bio and medical tribology are gaining ground for the development of new products in mechanics, chemistry, electronics, life sciences, and medicine.

This book aims to provide the fundamentals and advances in tribology for modern industry. Chapter 1 provides information on surface topography and chapter 2 is dedicated to basic aspects of friction and wear. Chapter 3 describes the fundamental aspects of lubrication and the relationship between lubrication and roughness. Chapter 4 contains information on micro and nano tribology while chapter 5 is dedicated to tribology in manufacturing. Finally, chapter 6 is dedicated to bio and medical tribology.

The book can be used as a textbook for the final undergraduate engineering course or as a topic on tribology at the postgraduate level. Also, it can serve as a useful reference for academics; tribology and materials researchers; mechanical, materials and physics engineers; and professionals in tribology and related industries. The scientific interest in this book will be evident for many important centres of research, including laboratories and universities throughout the world.

Therefore, it is hoped this book will inspire and enthuse other researches in this field.

The Editor acknowledges Woodhead/Chandos for this opportunity and for their enthusiastic and professional support. Finally, I would like to thank all the chapter authors for their contributions to this work.

J. Paulo Davim
University of Aveiro, Portugal
May 2010

List of figures

List of tables

About the contributors

Editor

J. Paulo Davim is an Aggregate Professor in the Department of Mechanical Engineering of the University of Aveiro, Portugal, and is Head of MACTRIB (Machining and Tribology Research Group). His main research interests include tribology/surface engineering and machining/manufacturing. He is the Editor in Chief of several international journals, Guest Editor of journals, book Editor, book Series Editor, and Scientific Advisor for many international journals and conferences.

Authors

Saverio Affatato is a Senior Research Scientist at the Istituto Ortopedico Rizzoli (IOR) in Bologna, Italy. In particular, he is responsible for the tribology area in the Laboratorio di Tecnologia Medica of the IOR. His main research interests include wear evaluation on hip and knee joint simulators and particle debris characterization. He is Referee of the international journals *Clinical Biomechanics*, *Acta Biomaterialia*, *Biomaterials*, and *Proc. IMechE Part H*.

María-Dolores Bermúdez is Head of the Materials Science and Engineering Research Group of the Materials and Manufacturing Department at the Technical University of

Cartagena, Spain. Her main research lines are currently focused on the study of tribological performance and surface interactions of materials using ordered fluids such as ionic liquids and the development of new composite materials with enhanced tribological performance using nanophases.

Leonid Burstein is based at Technion, Quality Assurance and Reliability Department, and at the Braude ORT College, Computer Engineering Department, Israel. His main research interests include hydrodynamic lubrication of roughened surfaces and system modelling. He is author of chapters in published scientific books and is an Editorial Board member and Reviewer for a number of international scientific periodicals. His achievements have also been reported in more than 60 publications in leading scientific journals.

Mark J. Jackson is Associate Department Head for Research and University Faculty Scholar at Purdue University, Indiana, USA. He is Director of the Advanced Manufacturing Laboratory and Leader of the Physics and Chemistry of Machining Group.

Ana-Eva Jiménez is research assistant and member of the Materials Science and Engineering Research Group of the Materials and Manufacturing Department at the Technical University of Cartagena, Spain. She is currently working on the study of the tribology and surface engineering of ionic liquids in contact with light alloys and high temperature materials.

Jonathan S. Morrell is Compatibility and Surveillance Manager at the Y12 National Security Complex, Oak Ridge, Tennessee, USA. Dr Morrell is Adjunct Professor at Purdue University and his research involves investigating the machining of pyrophoric materials.

Kausala Mylvaganam is a Visiting Research Fellow at the School of Mechanical and Manufacturing Engineering, University of New South Wales (UNSW), Australia. She received her PhD from the University of Cambridge, UK for her work on the *ab-initio* calculation of molecular properties. Currently she is doing research in nanotechnology with a particular focus on the modelling of ultra-precision machining and characterization of materials.

Prasanta Sahoo is a Professor in the Department of Mechanical Engineering, Jadavpur University, Kolkata, India. His main research interests include tribology and structural mechanics. He has authored a textbook on Engineering Tribology and a number of book chapters. He has co-authored more than 150 technical papers. He is the Associate Editor of one international journal and on the editorial board of five international journals.

Francesco Traina is a Medical Doctor in the Department of Traumatologia e Chirurgia Protesica e dei Reimpianti di Anca e di Ginocchio at the Istituto Ortopedico Rizzoli in Bologna, Italy. His main research interests include hip and knee implants, hip, knee, and ankle arthroscopy, and biological reconstruction of ligaments. He is Referee of the international journals *COOR*, *J Bone Jt Surg Br* and *Hip International.*

Liangchi Zhang is Scientia Professor, Australian Professorial Fellow and Professor of Mechanical Engineering at the School of Mechanical and Manufacturing Engineering, University of New South Wales (UNSW), Australia. He is an elected Fellow of the Australian Academy of Technological Sciences and Engineering. His research is in the field of precision and nano processing technologies, focusing on nanomechanics and nanomaterials, machining and solid mechanics.

1

Surface topography

P. Sahoo,

Jadavpur University, India

Abstract: This chapter discusses the approaches to solid surface topography characterization including the surface layers, roughness parameters and statistical aspects. The multiscale characterization of surface topography in terms of fractal analysis, Fourier transform and wavelet transformation is also considered. The measurement techniques for surface roughness evaluation are discussed in terms of surface profilometry, optical methods and electron microscopy including the advanced techniques like scanning tunnelling microscopy and atomic force microscopy.

Keywords: surface layers, roughness parameters, multiscale characterization, measurement techniques.

1.1 Introduction

Surface interactions are dependent both on the contacting materials and the shape of the surface. The shape of the surface of an engineering material is a function of both its production process and the nature of the parent material (Bhushan, 1996; Thomas, 1982; Whitehouse, 1994). When studied carefully on a very fine scale, all solid surfaces are found to be rough, the roughness being characterized by asperities of varying amplitudes and spacing. The distribution of the asperities are found to be directional when the finishing process is direction dependent, such as turning, milling, etc., and homogeneous for a non-directional finishing process like lapping, electro-polishing, etc. For the study of tribological behaviour it is essential to know the methods of measuring and describing the surface shape in general and the surface roughness in particular. The surface texture may include (a) roughness (nano- and micro-roughness), (b) waviness (macro-roughness), (c) lay and (d) flaw. Figure 1.1 shows a display of surface texture with uni-directional lay. Roughness is produced by fluctuations of short wavelengths characterized by asperities (local maxima) and valleys (local minima) of varying amplitudes and spacing. This includes the features intrinsic to the production process. Waviness is the surface irregularities of longer wavelengths and may result from such factors as machine or work piece deflections, vibration, chatter, heat treatment or warping strains. Lay is the principal direction of the predominant surface pattern, usually determined by the production process. Flaws are unexpected and unintentional interruptions in the texture. Apart from these, the surface may contain large deviations from nominal shape of very large wavelength, which is known as error of form. These are not considered as part of surface texture.

A very general typology of a solid surface is shown in Fig. 1.2. Deterministic surface textures may be studied by simple analytical methods. However, for most engineering surfaces, the textures are random, either isotropic or anisotropic, and either Gaussian or non-Gaussian; the exact type depends on the nature of the processing technique. So called cumulative processes such as peening, lapping and electro polishing where the final shape of each region is the

Figure 1.1 Display of surface texture

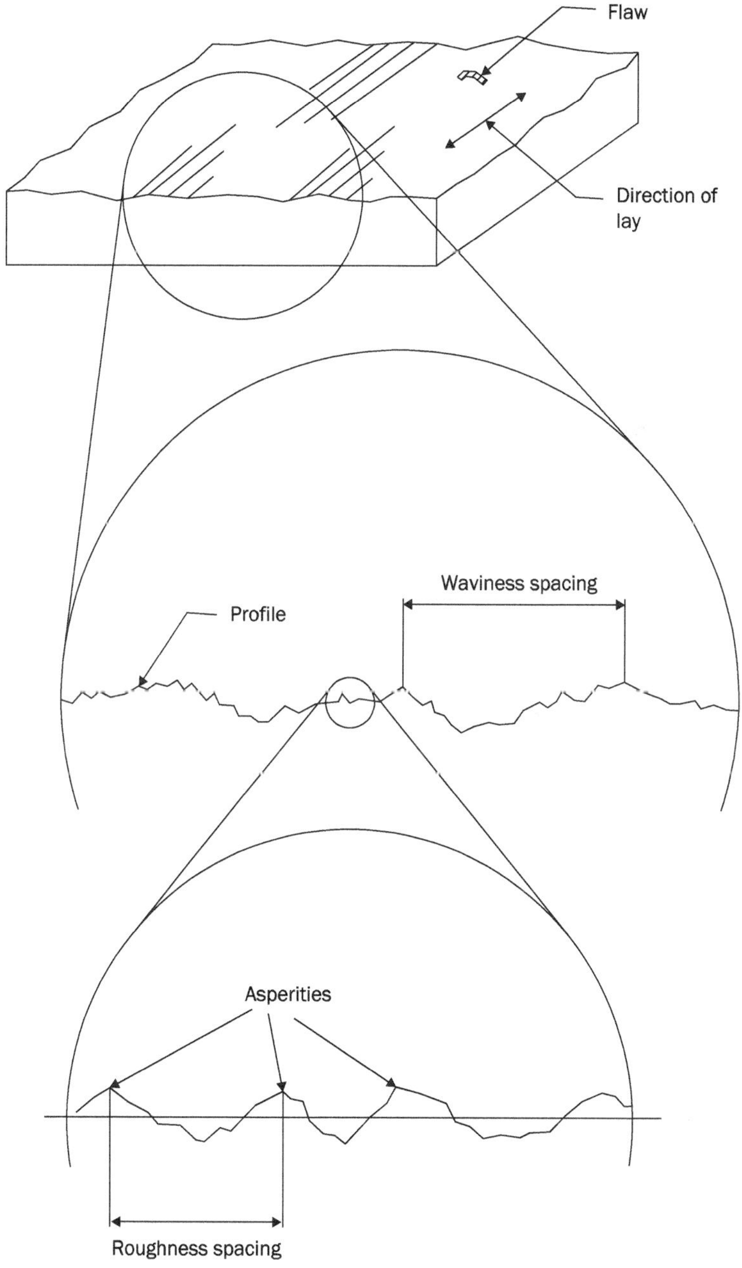

Figure 1.2 General typology of surfaces

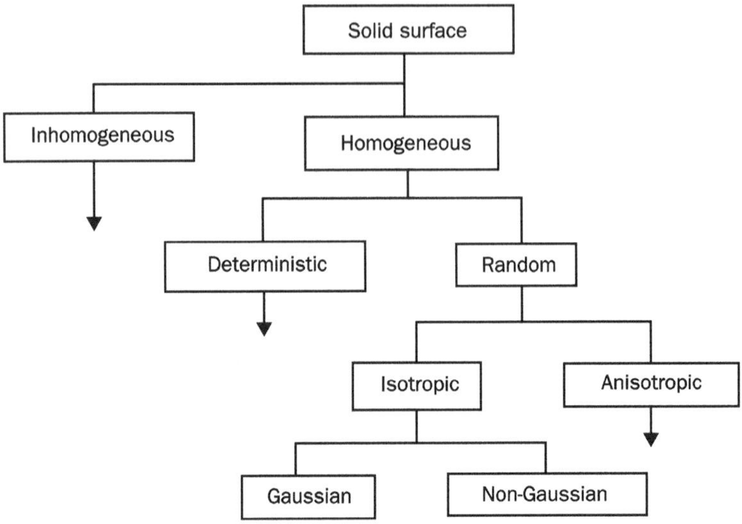

cumulative outcome of a large number of random discrete local events and independent of the distribution governing each individual event, produce surfaces that are governed by the Gaussian form. It is a direct result of the central limit theorem of statistical theory. Extreme-value processes such as grinding and milling and single-point processes such as turning and shaping usually produce anisotropic and non-Gaussian surfaces.

1.2 Characteristics of surface layers

The surface of a solid body is the geometrical boundary between the solid and the environment. But in tribological terms, surface includes the near-surface material to a significant depth. The surface of a typical metal consists of several layers whose physio-chemical properties are significantly different from that of the bulk material (Buckley,

1981). Such a typical metal surface with different layers is shown in Fig. 1.3. The top layer known as the Bielby layer, results from the melting and surface flow during the machining of molecular layers that are subsequently hardened by quenching as they are deposited on the cool underlying material. The layer is of amorphous or microcrystalline structure and thickness typically ranges from 1 to 100 nm. This is followed by a compound oxide layer, which is produced from the chemical reaction of the metal with the environment. Besides this, there may be absorbed films that are produced either by physisorption or chemisorption of oxygen, water vapour and hydrocarbons. In physisorption, no exchange of electrons takes place between the molecules of the absorbent and the absorbate. This involves van der Waals forces. In chemisorption, an actual sharing of electrons or electron interchange occurs between the chemisorbed species and the solid surfaces, and the solid surface bonds very strongly to the adsorption species through covalent bonds. The chemisorption layer is always monomolecular while physisorbed layers may be monomolecular or polymolecular. Heat of absorption for chemisorption (10 to

Figure 1.3 Typical surface layers

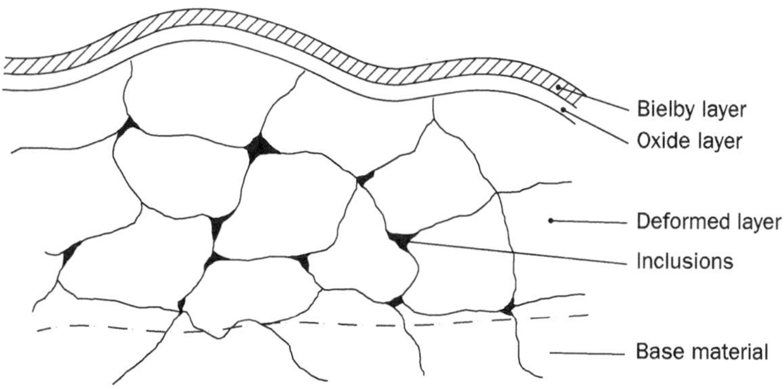

100 kcal/mol) is more than that for physisorption (1 to 2 kcal/mol) and chemisorption requires certain activation energy while physisorption needs no such energy. The thickness of oxide and chemically reacted layer ranges from 10 to 100 nm. Below this lies the deformed layer of the material containing some entrapped lubricants and contaminants followed by the bulk material. The thickness of the deformed layer ranges from 1 to 100 microns.

The tendency of molecules to absorb on the surface and the chemical reactivity may be regarded as extrinsic properties of the surface. The important intrinsic property of the surface is the surface tension or free surface energy, which is basically the reversible work required to create a unit area of the surface at constant volume, temperature and chemical potential. The creation of a new surface implies not only mechanical work but also heat consumption if the process occurs isothermally. The value of the surface energy of a material depends on the nature of the medium on the other side of the material boundary. Numerous surface analytical techniques are commercially available for the characterization of surface layers. The metallurgical properties like grain structure of the deformed layer can be obtained by sectioning the surface and examining the cross-section with the help of a high-resolution optical microscope or a scanning electron microscope (SEM). A transmission electron microscope (TEM) can be used to study microcrystalline structure and dislocation density. The crystalline structure of a surface layer can also be studied by X-ray, high-energy or low-energy electron diffraction techniques. An elemental analysis of a surface layer can be done with the help of an X-ray energy dispersive analyser (X-REDA), an Auger electron spectroscope (AES), or an electron probe microanalyser (EPMA), etc. The chemical analysis of the surface layers can be performed by X-ray photoelectron spectroscopy (XPS) and secondary ion mass spectroscopy (SIMS). Thickness and severity of the deformed layer can be obtained by measuring residual stress in the surface, while the thickness of all layers can be measured by depth-profiling a surface. The most common techniques for measurement of organic layer thickness include depth-profiling using XPS and ellipsometry.

1.3 Roughness parameters

Surface roughness basically refers to the variations in the height of the surface relative to a reference plane. It is in general measured either along a single line profile or along a set of parallel line profiles as in the case of a surface map. A surface is composed of a large number of length scales of superimposed roughness that are generally characterized by three different types of roughness parameters, viz., amplitude parameters, spacing parameters and hybrid parameters. Amplitude parameters are measures of the vertical characteristics of the surface deviations and examples of such parameters are centre line average roughness, root mean square roughness, skewness, kurtosis and peak-to-valley height. Spacing parameters are measures of the horizontal characteristics of the surface deviations and examples of such parameters are mean line peak spacing, high spot count, peak count, etc. On the other hand, hybrid parameters are a combination of both the vertical and horizontal characteristics of the surface deviations and examples of such parameters are root mean square slope of profile, root mean square wavelength, core roughness depth, reduced peak height, valley depth, material ratio, peak area and valley area. Hybrid parameters are considered more powerful than a parameter solely based on amplitude or spacing to characterize the surface topography.

Roughness is usually characterized by either of the two statistical height descriptors advocated by the International Standardization Organization (ISO) and the American National Standards Institute (ANSI) (Anonymous, 1985). These are CLA (Centre-line average, R_a) and RMS (Root mean square, R_q). Two other statistical height descriptors are rarely used – skewness (Sk) and kurtosis (K).

1.3.1 Centre line average (CLA)

It is defined as the arithmetic mean deviation of the surface height from the mean line through the profile. It is also termed as average roughness (symbol R_a). Here the mean line is defined so as to have equal areas of the profile above and below it. It may also be defined by the equation

$$R_a = \frac{1}{L}\int_0^L |Z(x)|dx \qquad [1.1]$$

where $Z(x)$ is the height of the surface above the mean line at a distance x from the origin and L is the measurement length of the profile (Fig. 1.4). The R_a value of a surface profile depends on its manufacturing method and some typical R_a (μm) values are: rough casting – 10, coarse machining – 3 to 10, fine machining – 1 to 3, grinding and polishing – 0.2

Figure 1.4 Centre line average of a surface over sampling length *L*

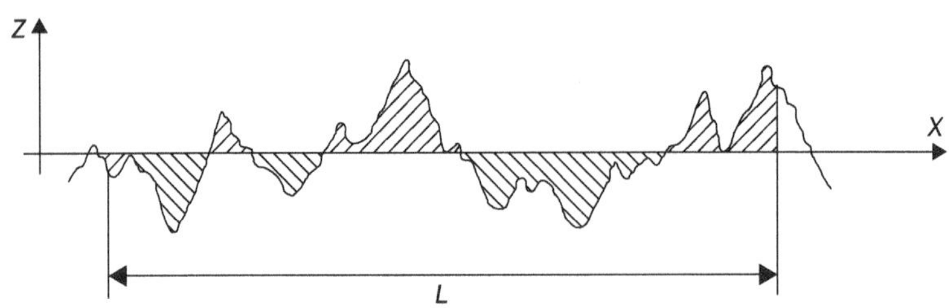

to 1 and lapping – 0.02 to 0.4. The disadvantage of using the R_a value is that this fails to distinguish between a sharp spiky profile and a gently wavy profile. It is possible for surfaces of widely varying profiles with different frequencies and shapes to have the same R_a value (Fig. 1.5).

Figure 1.5 Various surface profiles having the same R_a value

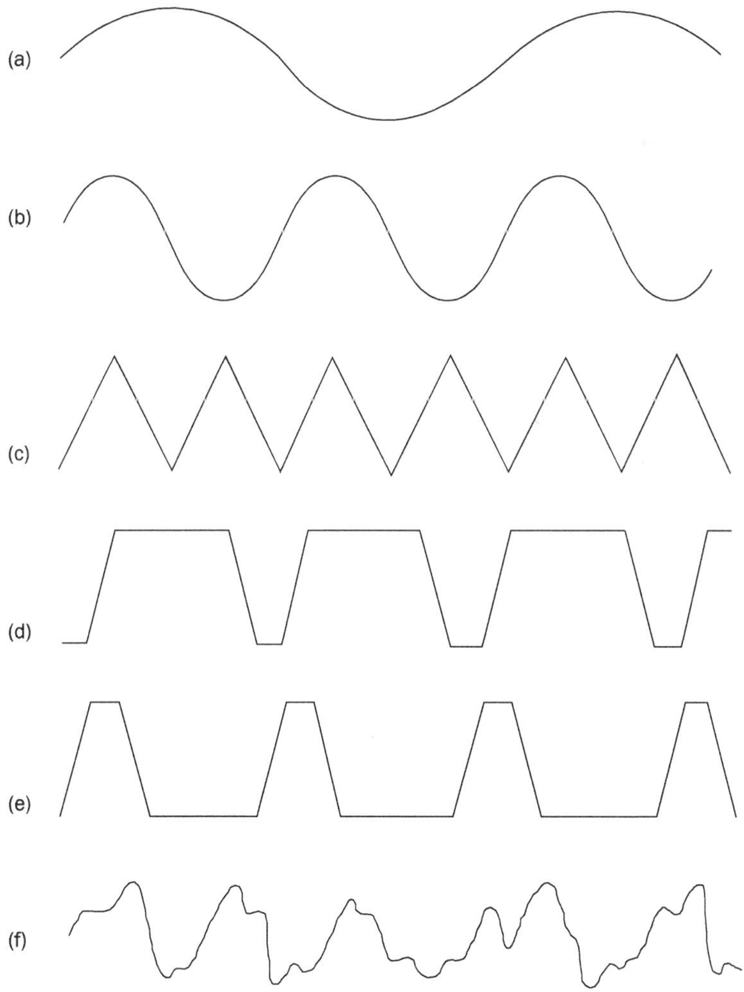

1.3.2 RMS roughness

This parameter represents the standard deviation of the distribution of the surface heights, so it is an important parameter to describe the surface roughness by statistical methods. This parameter is more sensitive than the arithmetic average height (R_a) to large deviation from the mean line. It is defined as the root mean square deviation of the profile from the mean line. It is denoted by the symbol R_q. The mathematical definition and the digital implementation of this parameter are as follows:

$$R_q = \sqrt{\frac{1}{L}\int_0^L [Z(x)]^2 dx} \quad [1.2]$$

The RMS mean line is the line that divides the profile so that the sum of squares of the deviations of the profile height from it is equal to zero.

1.3.3 Skewness and kurtosis

The skewness is a measure of the departure of a distribution curve from its symmetry and kurtosis is the measure of the bump on a distribution curve. The skewness and kurtosis in the normalized form may also be given as

$$Sk = \frac{1}{\sigma^3 L}\int_0^L Z^3 dx \quad [1.3]$$

and

$$K = \frac{1}{\sigma^4 L}\int_0^L Z^4 dx \quad [1.4]$$

where σ is the standard deviation of the distribution of asperity heights.

Some more extreme value height descriptors are also used as defined in Table 1.1.

Table 1.1 **Definitions of a few surface roughness parameters**

Symbol	Name	Definition
R_t	Peak-to-valley height	Separation of highest peak and lowest valley
R_p	Peak-to-mean height	Separation of highest peak and mean line
R_v	Mean-to-valley height	Separation of mean line and lowest valley
R_z (DIN)	Average peak-to-valley height	Average of single R_t values over five adjoining sampling lengths
R_z (ISO)	Ten point height	Separation of average of five highest peaks and five lowest valleys within single sampling length
R_{pm}	Average peak-to-mean height	Separation of average of five highest asperities and mean line

1.4 Statistical aspects

Another way of statistical treatment of a surface profile is to consider the probability distribution function of the height Z. The same is denoted by $P(Z)$ or $\phi(Z)$ and is obtained by plotting the number of occurrences of a particular value of Z in the data against the value of Z and normalizing the best fit curve to the data so that the total area enclosed by the distribution curve is unity. Thus it is given as

$$\int_{-\infty}^{+\infty} \phi(Z)dZ = 1 \quad [1.5]$$

The distribution function for most real surfaces is generally in the form of a 'bell-shaped' curve and can be described approximately by a Gaussian distribution given as

$$\phi(Z) = \frac{1}{\sigma\sqrt{2\pi}} \exp(-Z^2/2\sigma^2) \qquad [1.6]$$

where σ is the standard deviation of the distribution.

The shape of the distribution function may be quantified by means of the moments of the distribution. The *n*th moment of the distribution m_n is defined as

$$m_n = \int_{-\infty}^{+\infty} Z^n \phi(Z) dZ \qquad [1.7]$$

The first moment m_1 represents the mean line. The mean line is so located that m_1 is equal to zero. Then the second moment m_2 is equal to σ^2, the variance of the distribution. From definition of R_q it is seen that $R_q = \sigma$. It can also be shown that R_q/R_a for a Gaussian distribution comes out to be nearly 1.25. The third moment m_3 in normalized form gives the skewness, Sk $(= m_3/\sigma^3)$, which provides some measure of the departure of the distribution from symmetry. For a symmetrical distribution like Gaussian distribution, $Sk = 0$. The fourth moment m_4 in normalized form gives the kurtosis $(= m_4/\sigma^4)$, which is a measure of the sharpness of the peak of the distribution curve. For Gaussian distribution, $K = 3$. $K > 3$ means peak sharper than Gaussian and vice versa. Figure 1.6 shows a Gaussian distribution function as well as distribution functions with various skewness and kurtosis values, while Fig. 1.7 shows examples of surfaces with different skewness and kurtosis values. A surface with a Gaussian distribution has peaks and valleys distributed evenly about the mean:

- A surface with positive value of skewness has a wider range of peak heights that are higher than the mean.
- A surface with negative value of skewness has more peaks with heights close to the mean as compared to a Gaussian distribution.
- A surface with very low kurtosis has more local asperities above the mean as compared to a Gaussian distribution.

Figure 1.6 (a) Probability density functions for random distribution with different skewness; (b) symmetrical distributions (zero skewness) with different kurtosis

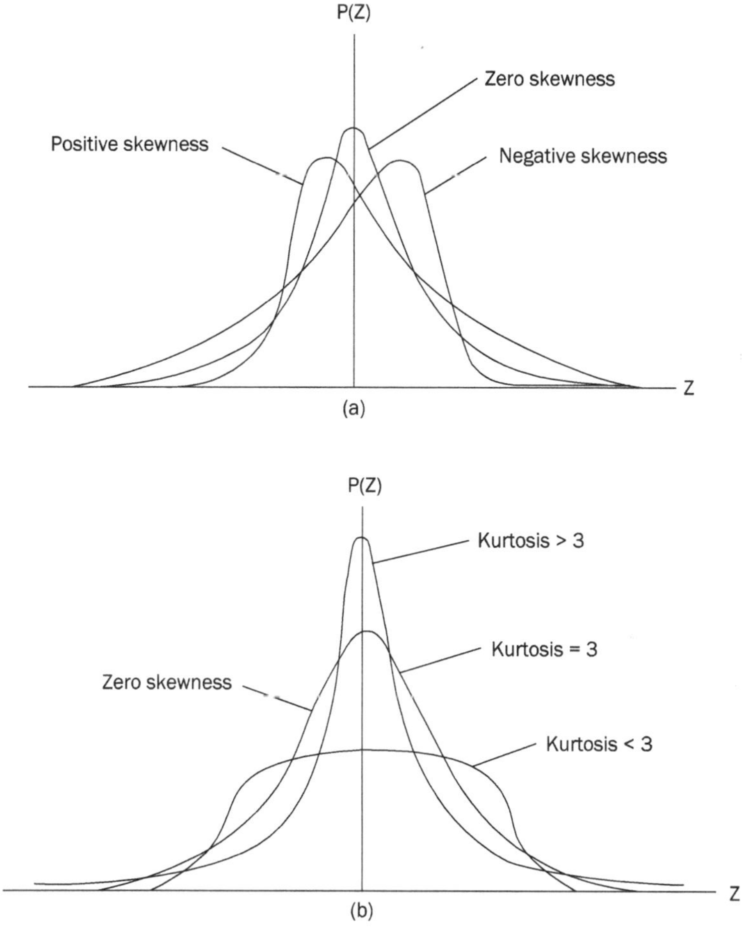

- A surface with very high kurtosis has fewer asperities above the mean.

In practice many engineering surfaces follow symmetrical Gaussian height distribution (Whitehouse, 1994). Generally, for most engineering surfaces the height distribution is

Figure 1.7 Schematic illustration for random functions with various skewness and kurtosis values

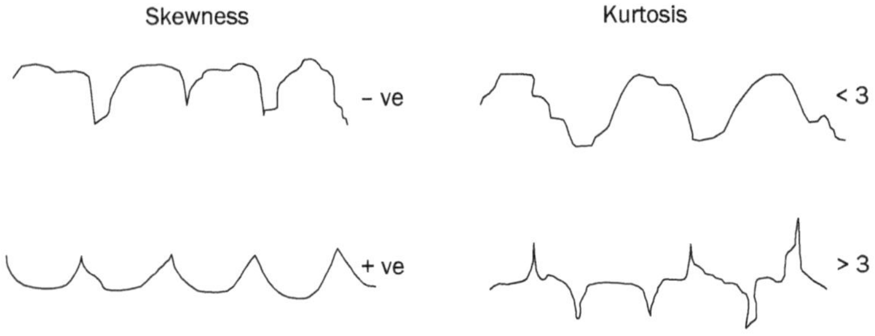

Gaussian at the high end and non-Gaussian at the lower end, the bottom 1–5% of the distribution (Williamson, 1968). Many common machining processes produce surfaces with non-Gaussian distribution: turning, shaping and electro discharge machining produce positively skewed surfaces; milling, honing, grinding and abrasion processes produce surfaces with negative skewness but high kurtosis values. Non-Gaussian surfaces are modelled using the well-known Weibull distribution and Pearson system of frequency curves.

For a digitized profile of length L with heights Z_i, $i=1$ to N, at a sampling interval $h=L/(N-1)$, where N represents the number of measurements, average height parameters are given as

$$R_a = \frac{1}{N}\sum_{i=1}^{N}|Z_i - m| \qquad [1.8]$$

$$\sigma^2 = \frac{1}{N}\sum_{i=1}^{N}(Z_i - m)^2 \qquad [1.9]$$

$$Sk = \frac{1}{\sigma^3 N}\sum_{i=1}^{N}(Z_i - m)^3 \qquad [1.10]$$

$$K = \frac{1}{\sigma^4 N}\sum_{i=1}^{N}(Z_i - m)^4 \qquad [1.11]$$

and

$$m = \frac{1}{N} \sum_{i=1}^{N} Z_i \qquad [1.12]$$

It is important to note here that all these statistical parameters are based on random data and hence they are subject to random statistical variations. They may not represent the true functional property of the surface in consideration. Moreover, none of these contain information on the horizontal or spatial distribution. There are a number of parameters that serve the description in spatial distribution. A few are described here.

1.4.1 Abbott bearing area curve

Sometimes it is required to estimate the proportion of the nominal area between two contacting surfaces that are in real contact. This is displayed on a curve known as the Abbott and Firestone bearing area curve (Abbott and Firestone, 1933). Figure 1.8 shows the construction of such a bearing area curve. A line parallel to the mean line is drawn at some height d and then the sum of all the intercepts along this line is expressed as a proportion of the total measurement length, given as $(a_1 + a_2 + \ldots)/L$. This actually gives a bearing

Figure 1.8 Construction of the Abbott bearing area curve from the topography of a surface: (a) surface profile; (b) bearing area curve

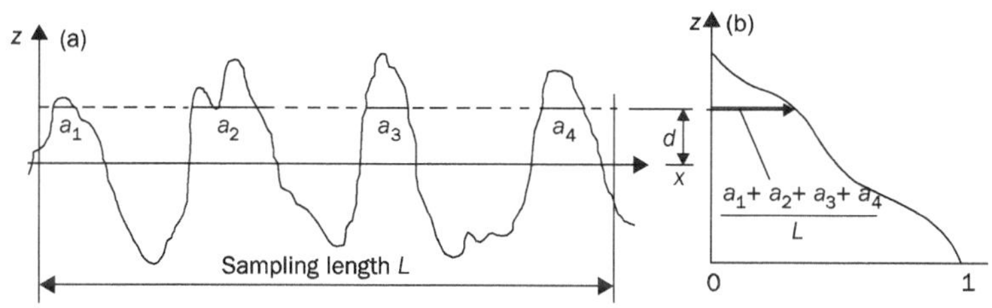

length. If the surface be isotropic, i.e., has no definite directional roughness, then the bearing length and bearing area are numerically equal.

1.4.2 Autocorrelation function (ACF)

The autocorrelation function (ACF) provides some information about the distribution of hills and valleys across the surface. The normalized ACF, $\rho(\beta)$, of a profile $Z(x)$ is defined as

$$\rho(\beta) = \frac{1}{\sigma^2}\left\{\lim_{L\to\infty}\frac{1}{L}\int_0^L Z(x).Z(x+\beta)dx\right\} \quad [1.13]$$

L being the sampling length and β the displacement along the surface (Fig. 1.9). When β is zero, the value of the normalized ACF $\rho(0)$ is a maximum and equal to unity. As β tends to infinity, the extent of correlation decreases and $\rho(\beta)$ tends to zero. If $\rho(\beta)$ is plotted against β, the curve decays from a value of unity to zero asymptotically at large values of β. For many real surfaces the ACF may be approximated by an exponential decay function. The form of the decay curve provides some

Figure 1.9 Graphical representation of the autocorrelation function

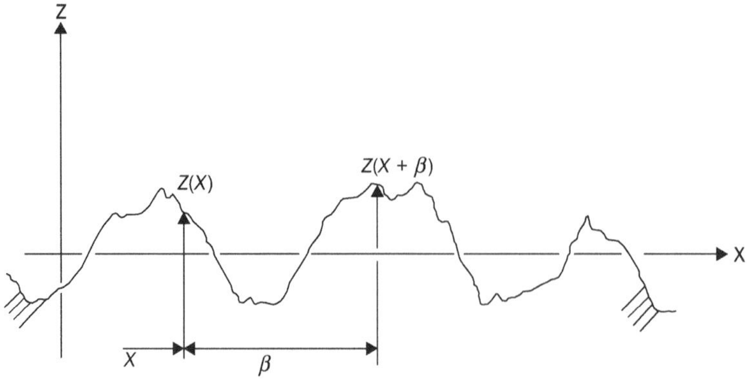

Figure 1.10 **Surface textures and their autocorrelation functions**

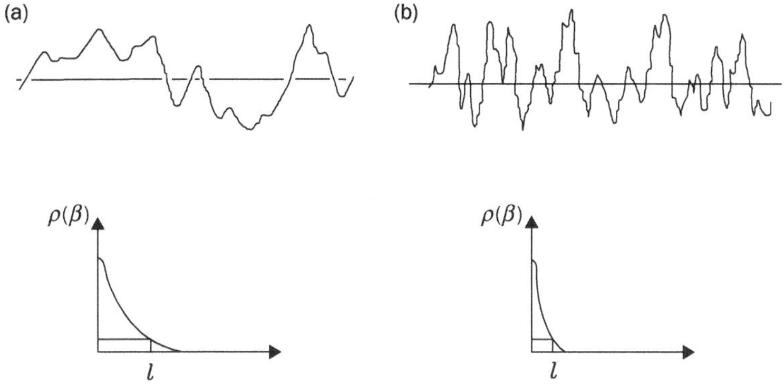

information on the horizontal distribution of roughness. Sometimes a correlation length l is defined as the value of β at which $\rho(\beta)$ equals 0.1. The value of this l is significantly higher in the case of an open texture surface than in a closed one (Fig. 1.10). It is suggested that the simple exponential decay function given by $\rho(\beta) = exp\ (-2.3\beta/l)$ is a good fit for many surfaces with randomness.

1.4.3 Power spectral density function (PSDF)

The power spectral density function $P(\omega)$ provides direct information about the spatial frequencies present in the profile. Particularly in the case of machined surfaces, it separates any strong surface periodicity resulting from the machining process. It is obtainable from the Fourier cosine transform of the autocorrelation function, given by

$$P(\omega) = \frac{2}{\pi}\int_0^\infty \rho(\beta)\cos(\omega\beta)d\beta \qquad [1.14]$$

In some modern profilometers, automatic computation of ACF and PSDF is possible.

1.5 Multiscale characterization of surface topography

The deviation of a surface from its mean plane is assumed to be a random process, which is characterized by the statistical parameters such as the variance of the height, the slope and the curvature. But, it has been observed that surface topography is a non-stationary random process. It means the variance of the height distribution is related to the sampling length and hence is not unique for a particular surface. Rough surfaces are also known to exhibit the feature of geometric self-similarity and self-affinity, by which similar appearances of the surface are seen under the various degrees of magnification as quantitatively shown in Fig. 1.11. Since increasing amounts of detail in the roughness are observed at decreasing length scale, the concepts of slope and curvature, which inherently assume the smoothness of the surface, cannot be defined. So the variances of slope and curvature depend strongly on the resolution of the roughness-measuring instrument or some other form of filter and are therefore not unique (Ling, 1990; Majumdar and Bhushan, 1990; Ganti and Bhushan, 1995; Sahoo and Roy Chowdhury, 1996). In contemporary literature such a large number of characterization parameters occur that the term 'parameter rash' is aptly used. The use of instrument-dependent parameters shows different values for the same

Figure 1.11 Qualitative description of statistical self-affinity for a surface profile

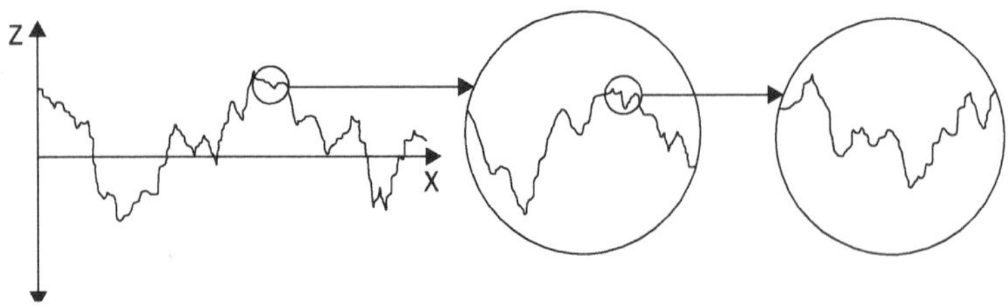

surface. Thus, it is necessary to characterize rough surfaces by intrinsic parameters, which are independent of all scales of roughness. Since this 'one-scale' characterization provided by statistical functions and parameters is insufficient to describe the multiscale nature of tribological surfaces, new 'multiscale' characterization methods need to be developed. Recent developments in this area have been concentrated mostly on four different approaches: Fourier transform methods; wavelet transformation methods; fractal methods; and the hybrid fractal-wavelet method.

Fourier transform methods basically decompose the surface data into complex exponential functions of different frequencies. The Fourier methods are used to calculate the power spectrum and the autocorrelation function in order to obtain the surface topography parameters. However, the difficulty with the application of these methods is that they provide results which strongly depend on the scale at which they are calculated, and hence they are not unique for a particular surface. This is because the Fourier transformation provides only the information whether a certain frequency component exists or not. As the result, the surface parameters calculated fail to provide information about the scale at which the particular frequency component appears.

Wavelet methods decompose the surface data into different frequency components and characterize it at each individual scale. While applying wavelets, the surfaces are first decomposed into roughness, waviness and form. Then the changes in surface peaks, pits and scratches, together with their locations, are obtained at different scales. However, there are still major problems in extracting the appropriate surface texture parameters from wavelets.

The fractal method incorporates fractal dimension which is an intrinsic property of multiscale roughness of surface characterization. It is invariant with length scales and is closely linked to the concept of geometric self-similarity. The self-similarity or self-affinity of rough surfaces implies that as the unit of measurement is continuously decreased, the surface area of the rough surface (a two-dimensional measure) tends to infinity and the volume (a three-dimensional measure) tends to zero. Here, self-similarity implies the property of equal magnification in all directions while self-affinity refers to unequal scaling in different directions. Thus, the Hausdorff or fractal dimension, $D + 1$, of rough surfaces is a fraction between 2 and 3. The profile of a rough surface $Z(x)$, typically obtained from stylus measurements, is assumed to be continuous even at the smallest scales. This assumption breaks down at atomic scale. But for engineering surfaces the continuum is assumed to exist down to the limit of a zero-length scale. Since repeated magnifications reveal the finer levels of detail, the tangent at any point cannot be defined. Thus the surface profile is continuous everywhere but non-differentiable at all points. This mathematical property of continuity, non-differentiability and self-affinity is satisfied by the Weierstrass-Mandelbrot (W-M) fractal function, which is thus used to characterize and simulate such profiles. The W-M function has a fractal dimension D, between 1 and 2, and is given by

$$Z(x)=L(G/L)^{D-1}\sum_{n=n_1}^{\infty}\frac{\cos 2\pi\gamma^{n}(x/L)}{\gamma^{(2-D)n}},\ 1<D<2,\ \gamma>1 \quad [1.15]$$

where G is a scaling constant. The parameter n_1 corresponds to the low cut-off frequency of the profile. Since surfaces are non-stationary random process the lowest cut-off frequency depends on the length L of the sample and is given by $\gamma^{n1} = 1/L$. The W-M function has the interesting mathematical property that the series for $Z(x)$ converges but that for dZ/dx diverges. It implies that it is non-differentiable at all points. The power spectrum of this W-M function can be expressed by a continuous function as

$$P(\omega) = \frac{G^{2(D-1)}}{2\ln \gamma_1} \frac{1}{\omega^{5-2D}} \qquad [1.16]$$

When this equation is compared with the power spectrum of a surface, the dimension D is related to the slope of the spectrum on a log-log plot against ω. The constant G is the roughness parameter of a surface, which is invariant with respect to all frequencies of roughness and determines the position of a spectrum along the power axis. In this characterization method both G and D are independent of the roughness scales of the surface and hence intrinsic properties. The constants of the W-M function, G, D, and n_1, form a complete and fundamental set of scale-independent parameters to characterize a rough surface.

The drawback of fractals is that they characterize surfaces at all scales while the wavelets provide a description at any particular scale. A possible solution to these problems is to use a combination of fractals and wavelets. Recently, a hybrid fractal-wavelet technique, based on the combination of fractal and wavelet methods, has been developed allowing for the 3-D characterization of often complex tribological surfaces with a unique precision and accuracy, without the need for any parameters. First, the surface topography features are broken down into individual scale components by wavelets and then fractals are applied to provide a surface topography description over the finest achievable range of scales.

1.6 Surface roughness measurement

The surfaces of any engineering component contain a vast number of peaks and valleys and it is not possible to measure the height and location of each of the peaks. So measurements are taken from a small and representative sample of the surface so chosen that there is a high probability for the surface lying outside the sample to be statistically similar to that lying within the sample. Over the years many different methods have been devised to study the topography of surfaces and a brief outline of some of them is presented here.

1.6.1 Surface profilometer

The most common method of studying surface texture features is the stylus profilometer, the essential features of which are illustrated in Fig. 1.12. A fine, very lightly loaded stylus is dragged smoothly at a constant speed across the surface under examination. The transducer produces an electrical signal, proportional to displacement of the stylus, which is amplified and fed to a chart recorder that provides a magnified view of the original profile. But this graphical representation differs from the actual surface profile because of difference in magnifications employed in vertical and

Figure 1.12 Component parts of a typical stylus surface-measuring instrument

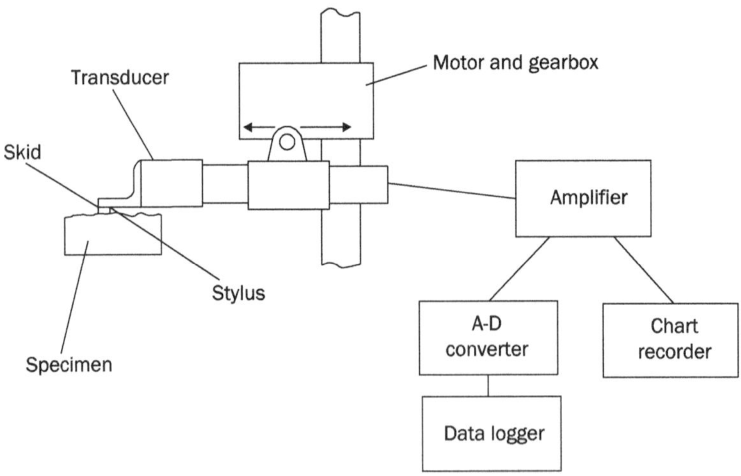

horizontal directions. Surface slopes appear very steep on a profilometric record though they are rarely steeper than 10° in actual cases. The shape of the stylus also plays a vital role in incorporating error in measurement. The finite tip radius (typically 1 to 2.5 microns for a diamond stylus) and the included angle (of about 60° for pyramidal or conical shape) results in preventing the stylus from penetrating fully into deep and narrow valleys of the surface and thus some smoothing of the profile is done. Some error is also introduced by the stylus in terms of distortion or damage of a very delicate surface because of the load applied on it. In such cases a non-contacting optical profilometer having optical heads replacing the stylus may be used. Reflection of infrared radiation from the surface is recorded by arrays of photodiodes and analysis of these in a microprocessor results in the determination of the surface topography. Vertical resolution of the order of 0.1 nm is achievable though maximum height of measurement is limited to a few microns. This method is clearly advantageous in cases of very fine surface features.

1.6.2 Optical microscopy

In this method, the surface of interest is held to reflect a beam of visible light and then these are collected by the objective of the optical microscope. An image of the surface is produced and is analysed at very high rates of resolution (up to 0.01 microns) by optical interferometers. Depth of field achievable is up to 5 microns. But success of the method depends on the reflective property of the material, which limits its use.

Optical methods may be divided into two groups: geometrical methods and physical methods. Geometrical methods include light-sectioning and taper-sectioning methods, while physical methods include specular reflection, diffuse reflection, speckle pattern and optical interference.

In the light-sectioning method, the image of a slit is thrown onto the surface at an incident angle of 45°. The reflected image appears as a straight line if the surface is smooth and as an undulating line if the surface is rough. In the taper-sectioning method, a section is cut through the surface to be examined at an angle of θ, thus effectively magnifying the height variation by a factor of $\cot\theta$, and is subsequently examined by an optical microscope. The surface is supported with an adherent coating that prevents smearing of the contour during the sectioning process, while the taper section is lapped, polished and lightly heat tinted to provide good contrast for optical examination. This process suffers from disadvantages like destruction of test surface and tedious specimen preparation.

In the specular reflection method, gloss or specular reflectance that is a surface property of the material and a function of reflective index and surface roughness, is measured by gloss meter. Surface roughness scatters the reflected light and affects the specular reflectance, thus a change in specular reflectance provides a measure for surface roughness.

The diffuse reflection method is particularly suitable for on-line roughness measurement during manufacture since it is continuous, fast, non-contacting and non-destructive. This method employs three varieties of approaches. In the total integrated scatter (TIS) approach, the total intensity of the diffusely scattered light is measured and used to generate the maps of asperities, defects and particles rather than micro-roughness distribution. The diffuseness of the scattered light (DSL) approach measures a parameter that characterizes the diffuseness of the scattered radiation pattern and relates this to surface roughness. In the angular distribution (AD) approach, the scattered light provides roughness height, average wavelength or average slope. With rougher surfaces, this may be useful as a comparator for monitoring both amplitude and wavelength surface properties.

In the speckle pattern method, surface roughness is related to speckle, which is basically the local intensity variation between neighbouring points in the reflected beam when a surface is illuminated with partially coherent light.

The optical interference technique involves looking at the interference fringes and characterizing the surface with suitable computer analysis. Common interferometers include the Nomarski polarization interferometer and Tolanski multiple beam interferometer.

1.7 Advanced techniques for surface topography evaluation

A further improvement in the resolution of surface topographic examination is possible by the use of electron microscopes. Two basic types of electron microscopes are available: scanning electron microscopes and transmission electron microscopes. In scanning electron microscopy (SEM) a focused beam of high-energy electrons is incident on the surface at a point resulting in the emission of secondary electrons. These are then collected and fed to an amplifier to send an electric signal to a cathode ray tube (CRT). The electron beam is scanned over the surface to have a complete picture and the CRT screen gives a topographical image of the entire area of interest. Depth of field is up to 1,000 microns, which acts as a primary advantage of this method over the optical method, but one drawback is the requirement on size of the specimen to be placed within the vacuum chamber of the instrument. This can be overcome by preparing a replica of the surface.

In transmission electron microscopy (TEM), the focused beam of high-energy electrons is made to transmit through a very thin specimen and the deflection and scattering of the electrons is recorded to analyse surface topography. Preparation of a specimen thin enough to transmit electrons plays a vital role and sometimes a replica of the surface retaining all the texture features but of a material having greater electron transparency is produced for the same purpose.

Recently, a different type of electron microscopy called scanning tunnelling electron microscopy (STM) has been introduced. It incorporates the electron-tunnelling phenomenon through an insulating layer separating two conductors. The sharp pointed tip of a probe forms one electrode and the surface of the specimen the other. The probe is moved by a highly precise positional controller to keep the tunnelling current at a steady value and provides an image of the surface under examination. The method is superior to the earlier ones in that it does not require any vacuum, but the one disadvantage is the poor design of the controller mechanism. The principle of the STM is very simple. Just like in a record player, the instrument uses a sharp needle, referred to as the tip, to investigate the shape of the surface, but the STM tip does not touch the surface. The schematic of the method is shown in Fig. 1.13. A voltage is applied between the metallic tip and the specimen, typically ranging between a few milli-volts and several volts. The tip touching the surface of the specimen results in a current and when the tip is far away from the surface, the current is zero. The STM operates in the regime of extremely small distances between the tip and the surface of only 0.5 to 1.0 nm, which are typically 2 to 4 atomic diameters. At these distances, the electrons can jump from the tip to the surface or vice versa. This jumping is necessarily a quantum mechanical process, known as 'tunnelling' and hence the name 'scanning tunnelling microscope'. The STMs usually operate at tunnelling currents between a few pico-Amperes (pA) and a few nano-Amperes (nA). The tunnelling current depends critically on the precise distance between the last atom of the

Figure 1.13 Schematic of working of STM

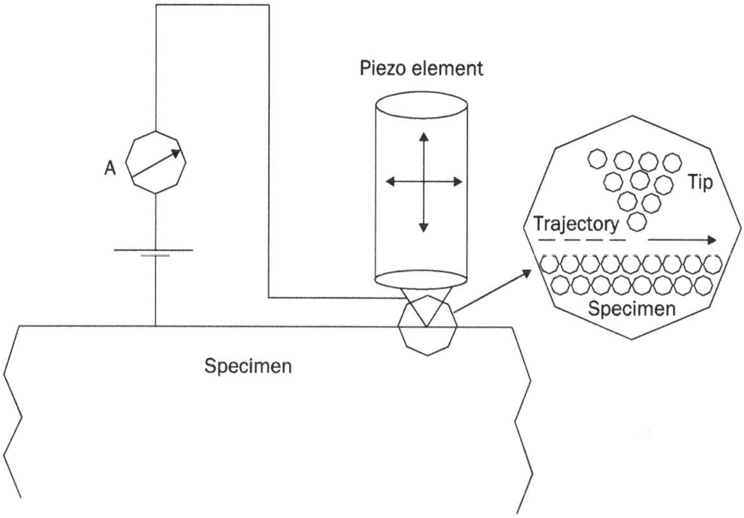

tip and the nearest atom or atoms of the underlying specimen. When this distance is increased a little bit, the tunnelling current decreases heavily. As a thumb rule, for each extra atom diameter that is added to the distance, the current becomes a factor of 1,000 lower. Thus the tunnelling current provides a highly sensitive measure of the distance between the tip and the surface. The STM tip is attached to a piezo-electric element, which changes its length a little bit when it is put under an electrical voltage. The distance between the tip and the surface can be regulated by adjusting the voltage on the piezo element. In most STMs, the voltage on the piezo elements is adjusted in a manner that the tunnelling current always has the same value, say 1 nA. Thus the distance between the last atom on the tip and the nearest atoms on the surface is kept constant. Using so called electronics, the distance regulation is done automatically. The feedback electronics continually measure the deviation of the tunnelling current from the desired value and accordingly adjust the position of the tip. While this feedback system is active, two other parts of the piezo elements are used to move the tip in a plane parallel to the surface to scan over the surface. In the scanning process, every time that the last atom of the tip is precisely over a surface atom, the tip needs to be retracted a little bit, while it has to be brought slightly closer when the tip atom is between the surface atoms. This automatically leads the tip to follow a bumpy trajectory, which replicates the atoms of the surface. Information about the trajectory is available in the form of the voltages that have been applied by the feedback electronics being visualized in the form of a collection of individual height lines or in the form of grey scale/ colour scale representation or in some three-dimensional perspective views.

More recently, the atomic force microscope (AFM) has been developed to investigate surfaces of both conductors and insulators on an atomic scale. Like the STM, the AFM relies on a scanning technique to produce very high-resolution, three-dimensional images of sample surfaces. In the AFM, the ultra-small forces (less than 1 nN) present between the AFM tip and sample surface are measured by the motion of a very flexible cantilever beam having an ultra small mass. The AFM combines the principles of the STM and the stylus profiler, but the important difference between the AFM and the STM is that in the AFM, the tip gently touches the surfaces. The AFM does not record the tunnelling current but the small force between the tip and the surface: the AFM tip is attached to a tiny leaf spring, the cantilever, which has a low spring constant, and the bending of the cantilever is detected with the use of a laser beam, which is reflected from the cantilever. The AFM thus measures contours of constant attractive or repulsive force. The detection is made very sensitive such that the forces as small as a few pico-Newton can be detected. Forces below 1 nano-Newton are usually sufficiently low to avoid damage to either the tip or the surface. Since AFM does not rely on the presence of a tunnelling current, it can also be used on non-conductive materials. Soon after the introduction of the AFM, it was realized that the same instrument could also be used to measure forces in the direction parallel to the surface, i.e., the friction forces. When modifications are incorporated for atomic scale and microscale studies of friction, it is termed as the friction force microscope (FFM) or the lateral force microscope (LFM). The FFM usually detects not only the deflection of the cantilever perpendicular to the surface, but also the torsion of the cantilever resulting from one lateral force. Schematic of the AFM/FFM commonly used for measurements of surface roughness, friction, adhesion, wear, scratching, indentation and boundary lubrication from micro to atomic scales is shown in Fig. 1.14.

In all surface profilometric methods, roughness (small-scale irregularities) and form error (deviation from its

Figure 1.14 Schematic operation of AFM/FFM

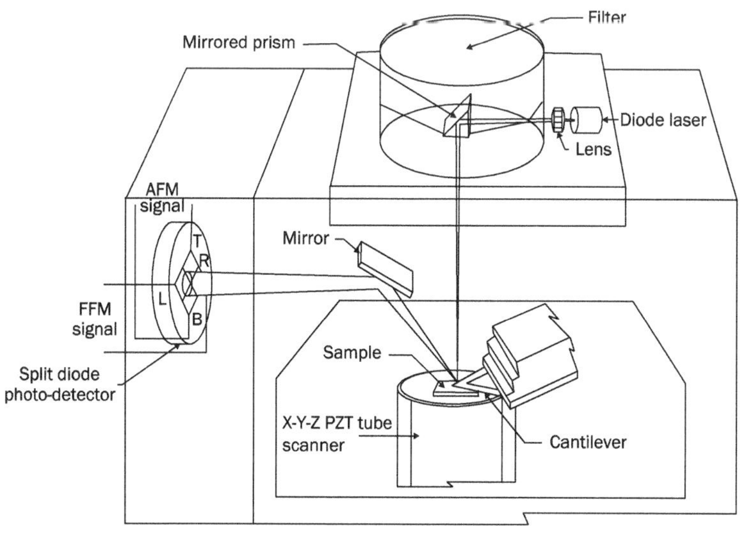

intended shape) remain coupled in the recorded data. Form error may be subtracted from the recorded data to provide only the roughness features by different means. The two most common methods used in stylus profilometer are the use of datum-generating attachments and the use of large radius skids or flat shoes. With these the average local level is used, as datum and form error or waviness is not recorded. Other methods include the use of filtering the displacement signal corresponding to waviness. Table 1.2 summarizes the comparison of the different roughness measuring methods.

1.8 Summary

Solid surfaces always contain deviations (roughness) from the prescribed geometrical form. Surface roughness is commonly characterized by average amplitude parameters: R_a and R_q. Height distribution and autocorrelation functions are used to completely characterize a random and isotropic surface. A surface contains a large number of length scales of roughness superimposed on each other, so commonly measured roughness parameters depend on the resolution of the measuring instrument and thus are not unique. Fractal analysis, Fourier transform and the wavelet transformation method can be used to characterize the multiscale nature of rough surfaces and various measurement techniques are used for roughness measurements. For on-line measurements, optical techniques such as specular reflection and diffuse reflection are commonly used. For off-line measurements, stylus profilers, atomic force microscopes and optical interferometers are very common.

Table 1.2 Summary and comparison of roughness measurement methods

Method	Quantitative data	3-D data	Resolution at maximum magnification (nm)		On-line measurement capability	Limitations/Comments
			Horizontal	Vertical		
Stylus method	Yes	Yes	15–100	0.1–1	No	Operates along linear track, contact type can damage sample, slow speed in 3-D mapping
Optical methods						
Light sectioning	Limited	Yes	500	0.1–1	No	Qualitative
Taper sectioning	Yes	No	500	25	No	Destructive, tedious specimen preparation
Specular reflection	No	No	10^5–10^6	0.1–1	Yes	Semiquantitative
Diffuse reflection	Limited	Yes	10^5–10^6	0.1–1	Yes	Smooth surfaces (<100 nm)
Speckle pattern	Limited	Yes			Yes	Smooth surfaces (<100 nm)
Optical interference	Yes	Yes	500–1000	0.1–1	No	
SEM	Limited	Yes	10	0.1	No	Operates *in vacuo*, limits on specimen size
TEM	Limited	Yes	0.5	0.02	No	Operates *in vacuo*, requires replication of surface
STM	Yes	Yes	0.2	0.02	No	Requires a conducting surface, scans small areas
AFM	Yes	Yes	0.2–1	0.02	No	Scans small areas

1.9 References

Abbott, E. J. and Firestone, F. A. (1933), 'Specifying surface quality', *Mech. Eng.* 55: 569–72.

Anonymous (1985), *Surface Texture (Surface Roughness, Waviness, and Lay)*, ANSI/ASME B46.1, ASME, New York.

Bhushan, B. (1996), *Tribology and Mechanics of Magnetic Storage Devices* (2nd edn), Springer, New York.

Buckley, D. H. (1981), *Surface Effects in Adhesion, Friction, Wear and Lubrication*, Elsevier, Amsterdam.

Ganti, S. and Bhushan, B. (1995), 'Generalized fractal analysis and its applications to engineering surfaces', *Wear* **180**: 17–34.

Ling, F. F. (1990), 'Fractals, engineering surfaces and tribology', *Wear* **136**: 141–56.

Majumdar, A. and Bhushan, B. (1990), 'Role of fractal geometry in roughness characterization and contact mechanics of surfaces', *ASME J. Tribology* **112**: 205–16.

Sahoo, P. and Roy Chowdhury, S. K. (1996), 'A fractal analysis of adhesion at the contact between rough solids', *Proc. IMechE. Part J: J. Engineering Tribology* **210**: 269–79.

Thomas, T. R. (1982), *Rough Surfaces*, Longman, London, UK.

Whitehouse, D. J. (1994), *Handbook of Surface Metrology*, Institute of Physics Publishing, Bristol, UK.

Williamson, J. B. P. (1968), 'Topography of solid surfaces', in *Interdisciplinary Approach to Friction and Wear* (P. M. Ku, ed.), SP-181, pp. 85–142, NASA Special Publication, NASA, Washington, DC.

2

Friction and wear

A.-E. Jiménez and M.-D. Bermúdez, Universidad Politécnica de Cartagena, Spain

Abstract: This chapter is an introduction to the dry friction and wear for sliding contacts. The concepts of friction and wear, their corresponding coefficients and their main mechanisms are described, including stick-slip effects, adhesion and ploughing. Adhesive, abrasive, erosive and erosion-corrosion wear mechanisms are described. Friction–wear relationships are discussed and wear maps are introduced.

Keywords: dry friction and wear, friction and wear mechanisms.

2.1 Friction

Friction between contacting bodies is manifested as the force that must be overcome to initiate or sustain motion, and as the energy dissipated during relative motion (Bogdanovich et al., 2009). The term friction describes the gradual loss of kinetic energy for materials or bodies in relative motion (Röder et al., 2000), while the dry friction of materials can be defined as the resistance to the relative movement between two surfaces in contact (*ASM Handbook*, vol. 18, 1992; Bushan, 2001; Czichos, 1978). From an engineering point of view, friction is a major cause of energy waste dissipated as heat, and a major cause of components and equipments failure (Bayer, 2002).

2.1.1 Friction coefficient, friction laws

Sliding friction has been described by the following laws (Hutchings, 1992; Williams, 1996):

First law of friction

The first law states that the friction force F between a pair of loaded sliding surfaces is proportional to the normal load W (Fig. 2.1), that is, the tangential force required to slide a body along a surface is proportional to the weight of the body. The proportionality constant between F and W is known as the coefficient of friction, μ.

Figure 2.1 Scheme of two contacting bodies in relative motion

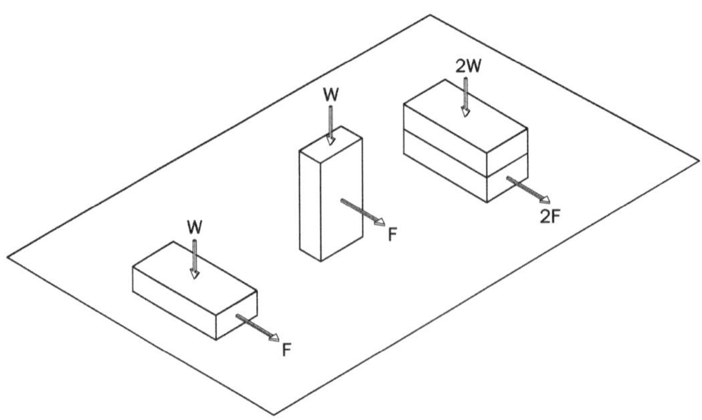

Coefficient of friction: Definition

The first law can be expressed by the simple equation:

$$F = \mu W \quad [2.1]$$

where μ is called the coefficient of friction. This expression also implies the second law.

Second law of friction

The second law of friction is derived from the first:

> The friction force is independent of the apparent area of contact.

Once sliding stops, the force needed to initiate sliding (static friction) is greater than the force needed to sustain sliding (kinetic friction). The dependence of friction force with sliding velocity is very small.

Third law of friction

> Kinetic friction is independent of sliding velocity, which is really an approximation.

The first two laws are known as Amontons' laws, who formulated them in the Proceedings of the French Academy of Sciences in 1699. The third law is due to Coulomb.

These laws are not derived from fundamental principles, but they are rather empirical ones based on experimental observations. In order to understand when these laws can be applied, it is necessary to analyse their physical origins (Santner et al., 2005).

2.1.2 Static and kinetic friction

When considering how friction originates at the atomic level, it is convenient to separate the problem into two regimes:

1. Static friction: the force needed to overcome the potential energy barriers between atoms in order to initiate sliding.
2. Kinetic friction: the mechanisms for dissipating energy as atoms slide over each other.

Figure 2.2 shows a friction curve as a function of sliding displacement. The maximum friction value appears after a short sliding distance from the origin.

2.1.3 Stick-slip

In practice, friction-distance records obtained in tribological tests are not smooth curves. Sliding surfaces often exhibit stick-slip behaviour rather than continuous forces (Berman et al., 1996; Chatelet et al., 2008; Rapoport, 2009; Yoshizawa et al., 2003).

Figure 2.3 shows the real-time variation of friction with sliding distance for a pin-on-disc metal-metal contact. After a maximum friction value during the running-in period, a

Figure 2.2 Variation of the friction coefficient with sliding distance

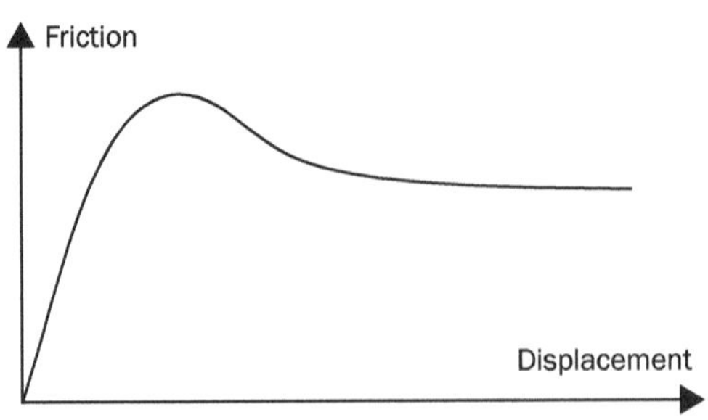

Figure 2.3 Friction coefficient–sliding distance record obtained in a pin-on-disc test for a metal–metal contact under dry conditions

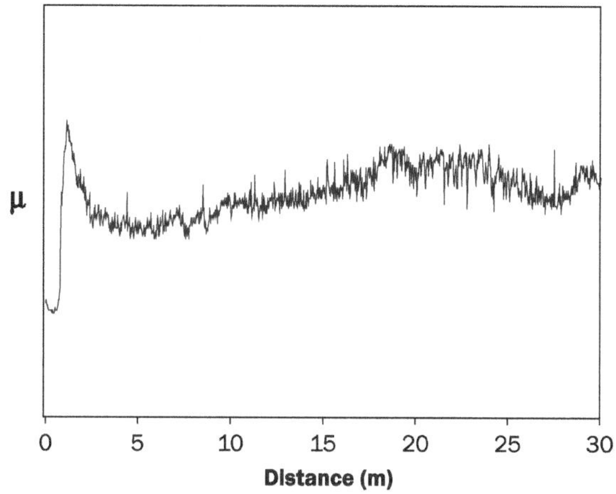

steady-state is reached, where oscillations of friction are observed due to the stick-slip effect.

Stick-slip, or interrupted motion rather than smooth uninterrupted motion (Fig. 2.4) occurs in many different phenomena such as friction, fluid flow, material fracture

Figure 2.4 Stick-slip effect

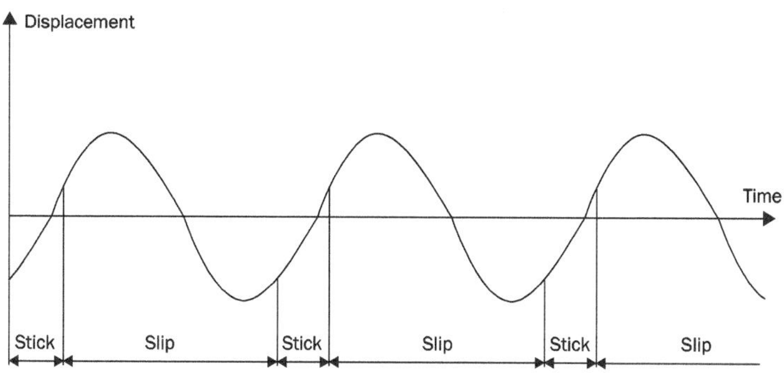

Adapted from Hutchings (1992).

and wear, sound generation, and sensory 'texture'. During stick-slip, a system is believed to undergo transitions between a static (solid-like) state and a kinetic (liquid-like) state.

Two smooth, cold, metal surfaces form tiny spot welds that have to be broken apart before they can slide over each other. This is another reason why metals stick as they slip if they are pressed together and pushed.

Such microscopic causes of friction and wear are increasingly important (Kim et al., 2009; Mate, 2008) as the mechanical engineering size scale decreases. Here, conventional methods of lubrication start to fail.

Roughness is thought to be behind most stick-slip. Even an apparently smooth sheet of metal or glass is usually covered with tiny ridges, pits and scratches. Figure 2.5 shows the topography of engineering materials surfaces. These surface

Figure 2.5 Three-dimensional topography map of a metal surface obtained by optical profilometry

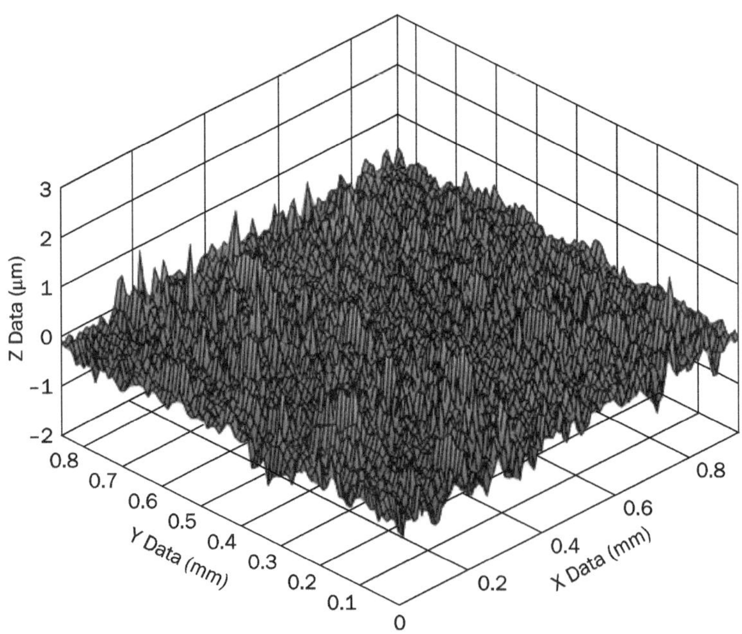

Figure 2.6 Contacts between the surface asperities

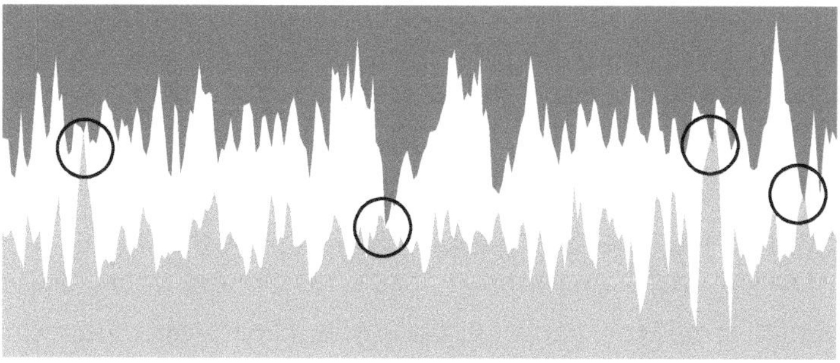

asperities (Fig. 2.6) interlock until the driving force is high enough to break the irregularities or slide them over one another. Relative motion does not occur until some critical shear stress is reached. In that moment, the interfaces slip each other to relieve the resultant stress. This occurs when the static frictional force is greater than the kinetic frictional force during sliding. Stick-slip behaviour depends on surface topography and on the elastic and plastic properties of the sliding materials.

For large objects sliding over one another, the friction force is proportional to the true contact area between the two bodies, which is smaller than the apparent contact area because the surfaces are rough, consisting of a large number of asperities that actually make the contact (Fig. 2.6).

The situation for nanomaterials, however, has been unclear, since the continuum contact theory that can account for macroscale effects has been predicted to break down at the nanoscale. Large-scale molecular dynamics simulations of scanning force microscopy experiments (Mo et al., 2009; Mate, 2008) show that, despite this, simple friction laws do apply at the nanoscale: the friction force depends linearly on the number of atoms, rather than the number of asperities, that are chemically interacting across the sliding interfaces.

2.1.4 Friction models

As we have seen (Fig. 2.5) engineering surfaces are never flat. Contact takes place through a number of surface asperities (Fig. 2.6) of different shape and size.

In unlubricated sliding between flat surfaces, friction is due to the elastic and plastic deformation processes that suffer the surface asperities in contact. Individual asperities would support a fraction w of the total normal force, W. The contact area for each junction (A_w) between individual asperities would be

$$A_w = w/H \qquad [2.2]$$

where H is the hardness of the softer material.

The real contact area results in

$$A = W/H \qquad [2.3]$$

Plastic deformation takes place when the applied shear pressure is higher than the shear strength (σ_s) of the surface junction. The tangential force under sliding is then

$$F_T = \sigma_s A \qquad [2.4]$$

and the friction coefficient μ, defined as the ratio between the tangential and the normal forces, is

$$\mu = \sigma_s/H \qquad [2.5]$$

Under these conditions, the dry friction coefficient is not dependent on the normal load or sliding velocity, but depends on the surface properties of the sliding materials and the specific conditions of operation. With the exception of hydrodynamic lubrication, the nominal range for the coefficient of friction for sliding is situated between 0.01 and 2 (Mate, 2008).

The effects of sliding speed, roughness and other surface and environmental conditions are generally much more pronounced and significant in engineering. Transitions in friction and the coefficient of friction are often the result of melting or thermal softening, or due to the formation of different layers on or between surfaces, such as oxides and tribofilms.

Expressions for the coefficient of friction, based on adhesion, deformation and hysteresis mechanisms, have been developed (Anderson et al., 2007; Blau, 2009b) to explain friction behaviour in general and for specific tribosystems. The friction force arises not only from the surface mechanical contact conditions, but also from energy loss, so that the mechanisms of friction can be divided into those associated with adhesion and those due to deformation effects.

The microscopic mechanisms (Hsu, 1996) that generate friction are: adhesion, mechanical interlocking of surface asperities, ploughing by surface asperities, deformation and fracture, plastic deformation by wear particles, and third bodies.

Adhesion and ploughing in friction

Adhesion

Two sliding surfaces are in contact at a certain number of sites (Fig 2.6), so that the real contact area is much smaller than the nominal area. These contact points support the whole load so the contact pressure is higher than the applied normal load.

As shown in Fig. 2.5, sliding surfaces are never completely smooth. Surface roughness derived from surface asperities is in the origin of friction through two main mechanisms:

1. The adhesion force needed to shear the contacting junctions where adhesion occurs.
2. The ploughing force needed to plough the asperities of the harder surface along the softer one.

Bowden and Tabor (1986) proposed that the interfacial friction under dry or poorly lubricated conditions can be determined by the minimum force *F* required to shear the welded junction formed by adhesion bonds between contacting asperities as shown in Fig. 2.6. The high friction coefficient found for clean metals is attributed to the growth of junctions resulting from the combined action of normal pressure and tangential load under dry and boundary lubricated conditions. Under these conditions, the load is supported by the interface asperities and the friction is controlled by the growth of the real contact areas.

The friction for two sliding surfaces with contact between the surface asperities can be related to elastic and plastic deformation. In this approximation, each contact between asperities would carry the corresponding fraction of the total normal load. The total contact area between the two bodies can be considered as the W/H ratio. The friction coefficient (μ) defined as in equation 1, can be calculated as the ratio between the shear strength and the hardness of the softer sliding material. In this model, the friction coefficient is not dependent on the normal applied load.

Two distinct friction regions can be distinguished. In the first region, the friction coefficient increases rapidly due to the transition from elastic to plastic deformation between the contacting asperities. In the second region, where the friction is dominated by ploughing, the friction coefficient remains constant. The transition between both friction regimes occurs when the contacting asperities deform plastically. The critical shear factor determines the ploughing regime.

The relationship between adhesive forces and friction forces takes place through the contacting asperities of the two surfaces, which undergo elastic and plastic deformation, with interatomic attractive and repulsive forces. When a tangential force is applied to slide one surface over the other, shear stresses develop over the junction interfaces to resist this force. At low shear stresses, the interatomic forces prevent the atoms from sliding, and the material around the contact deforms elastically. When a critical shear stress is reached, the applied force is greater than the interatomic forces, and sliding starts. If it is assumed that all junctions have the same shear strength σ_s, the adhesive friction force F_{adh} necessary to shear all the junctions and slide the object would be given by:

$$F_{adh} = A \cdot \sigma_s \qquad [2.6]$$

where A is the total real area of contact.

In practice, the shear strength is likely to vary from junction to junction, so σ_s corresponds to the average shear strength of the junctions. The real area of contact varies linearly with the load W. If the shear strength is independent of contact pressure, F_{adh} will also be independent of the apparent area of contact.

Most asperities on metal surfaces finished by grinding or polishing deform plastically during initial contact. For this situation,

$$A - W/H \qquad [2.7]$$

and

$$F_{adh} = \sigma_s W/H \qquad [2.8]$$

This finally leads to

$$\mu_{adh} = \sigma_s /H \qquad [2.9]$$

for shearing junctions where most of the deformation is plastic.

Ploughing

The contribution to friction force from ploughing hard asperities through a softer surface is called the ploughing friction or F_{plough}. A rigid, cone-shaped hard single asperity (Fig. 2.7) applies a ploughing force which depends on H (hardness of the soft material) and the cross-sectional area of the groove

$$ax = x^2 \tan\varphi \quad [2.10]$$

As the asperity ploughs through the softer material, the load (W) is supported by the contact pressure H of the softer material acting over an area of $\pi a^2/2$ below the conical indenter:

$$W = {}^1\!/\!_2(H\pi a^2) = {}^1\!/\!_2(H\pi x^2 \tan^2\varphi) \quad [2.11]$$

Thus, the ploughing friction coefficient would be

$$\mu_{plough} = F_{plough}/W = [(2/\pi)\cot\varphi] \quad [2.12]$$

As the adhesive and ploughing contributions to friction are not completely independent (Lovell and Deng, 2000), it is convenient to treat them separately and express the total friction force F as $F_{adh} + F_{plough}$. If this would be the case, the coefficient of friction should not exceed 0.2 for contacting metals with similar hardness, or 0.3 for a hard metal sliding on a softer one. However, experimental friction values are often much higher, thus indicating that other mechanisms

Figure 2.7 A hard conical asperity ploughing through a softer surface

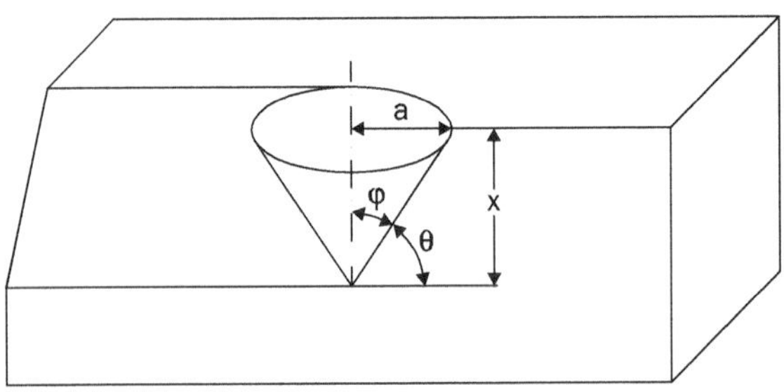

(Anderson et al., 2007; Blau, 2009a; Mate, 2008) contribute to friction, such as junction growth and work hardening.

2.1.5 Friction–wear relationship

There are three generic mechanisms usually proposed for friction: adhesion, deformation and hysteresis (Blau, 2009b; Kato, 2000). As we have seen, adhesion involves the shearing of junctions formed between two contacting surfaces, while deformation involves the displacement of material as a body moves across another. Finally, hysteresis refers to the response time necessary for a solid to react to the changes in the forces applied to it. This type of mechanism is particularly significant for viscoelastic materials, such as polymers. These friction mechanisms may produce wear, but not necessarily.

In the case of adhesion, wear does not result if shearing occurs at the interface. In the case of deformation, wear results directly from plastic deformation or cutting. Elastic deformation and hysteresis respond to the application of forces. However, neither of them cause wear in a single cycle.

Only a small portion of the energy dissipated by friction produces wear (it is estimated to be less than 10%). The remaining portion is mainly dissipated as heat (Bogdanovich et al., 2009), although other mechanisms also contribute, such as acoustic emission, changes in surface roughness, wear debris formation, tribochemical and microstructural processes.

Friction and wear are closely related but are distinct phenomena. Wear mechanisms contribute to friction, because wear processes require the application of force and consume energy. At the same time, wear mechanisms are affected by the shear loading resulting from friction and by the increase in temperature caused by frictional heating, so friction can influence wear behaviour. In addition, friction behaviour can be influenced by the changes to a surface caused by wear.

There is no general correlation between the coefficient of friction and the normalized wear rates. However, changes in the coefficient of friction with time or sliding distance are often associated with changes in wear behaviour. In some cases, a reduction in the coefficient of friction with continued sliding (transition from running-in to steady-state) can be associated with the formation of a stable transfer film.

The friction coefficient–wear rate relationship under sliding conditions is a function of the mechanism of kinetic energy conversion and dissipation processes. The same values of friction coefficient can be obtained for different wear mechanisms due to a different mechanism of friction work dissipation. More energy must be dissipated from the contact as the sliding frictional work increases.

2.2 Wear

Wear is the damage to a solid surface, generally involving progressive loss of material, due to the relative motion between that surface and a contacting material or substance. Wear is determined as the volume loss from solid surfaces in moving contact.

Despite the technological importance of wear, the multitude of physical mechanisms contributing to wear makes it difficult to develop a simple and universal model to describe it (Meng and Ludema, 1995; Rice and Moslehy, 1997).

The coefficients of friction and wear are not material properties but two kinds of responses of a tribo-system. They are always reasonably related with each other when the necessary functions of the tribo-system are well considered. Variations in friction and wear can be explained due to the effect of surface roughness, hardness, ductility, surface oxide films, tribolayers and transfer processes. For metals sliding against metals under unlubricated conditions, one or several of the following processes are observed: growth of oxide surface layers; moisture adsorption and surface layer fracture; plastic deformation and microstructural changes of surface layers; adhesive transfer and retransfer of material between the counterparts; detachment; agglomeration; compaction and mechanical milling of wear particles; changes in surface morphology and roughness at the contact.

2.2.1 Sliding wear: Archard equation

The most elementary definition of wear is the loss of material from one or both of the contacting surfaces when subjected to relative motion, while a wider definition includes any form of surface damage caused by rubbing one surface over another. Wear is a complex process, making it one of the more difficult aspects of tribology to study.

Despite the technological importance of wear, no simple and universal model has been developed to describe it. As with many other tribological phenomena, the multitude of physical mechanisms contributing to wear makes it difficult to develop a general understanding of wear at the nanoscale.

A simple model for sliding adhesive wear is due to Archard (1953) and is referred as the Archard wear equation. This model derives an equation for the wear rate Q as the volume V of material removed per unit sliding distance, d. A basic assumption of this wear model is that the wear volume is proportional to sliding distance. This is true as long as the wear mechanism does not change. Thus during the running-in or breaking-in period, the initial wear rate can be higher or lower than that found once the steady-state is reached.

Another assumption of this model is that wear occurs where surface asperities touch, so the true area of contact is the sum of all the individual asperity contact areas. For each individual asperity contact, the maximum contact area for an asperity of radius a, is πa^2.

Wear takes place when a fragment of material detaches from an asperity during contact. In this wear model, the volume of this fragment is presumed to be proportional to the cube of the contact dimension a, therefore the volume of this wear fragment can be approximated as $2/3\pi a^3$, for a sliding distance of $2a$.

However, only a fraction κ of the contacts generate volume loss. The wear volume per unit of sliding distance is

$$[(2/3\pi a^3)/2a] = \kappa(\Delta A/3) \qquad [2.13]$$

The wear rate is the total volume loss

$$Q = \kappa/3 = [\Sigma(\Delta A)] \qquad [2.14]$$

for the real area of contact between the two surfaces.

The Archard model considers that the deformation of the asperity contacts is predominantly plastic. This leads to the real contact area A being proportional to the load and described as

$$A = W/H \qquad [2.15]$$

where W = applied normal load; H = hardness of the material that suffers volume loss.

By defining the adimensional wear coefficient K as $\kappa/3$, the final equation results:

$$Q = KW/H \qquad [2.16]$$

Dry wear coefficient values are usually between 10^{-2} and 10^{-7}. The Archard equation is expressed as

$$Q = kW \qquad [2.17]$$

where k is K/H, and is called the dimensional wear coefficient or just wear coefficient expressed in units of $mm^3(Nm)^{-1}$. The wear coefficient compares the severity of the wear damage, but is not sufficient to account for the mechanisms which produce wear in each case.

In practice, wear coefficients can vary dramatically as the sliding conditions change. Transient wear rates occur during the initial or running-in period, while the sliding surfaces evolve towards their steady-state sliding conditions. For most sliding systems, no single wear mechanism operates, rather several mechanisms are present, but their relative importance changes as the sliding conditions change, with wear rate abruptly changing when different wear mechanisms predominate. Many variables influence the wear damage processes of metals. Wear maps have been represented to show how wear rates change with sliding conditions using dimensionless variables; the normalized contact pressure has been defined as the loading force divided by the nominal contact area and the hardness of the softer material; and the normalized velocity, the sliding velocity, divided by the velocity of heat flow.

2.2.2 Fretting wear

Fretting is defined as contact between surfaces subjected to reciprocating motion of low amplitude, while reciprocating wear occurs at much higher amplitudes. Surface degradation usually occurs when the amplitude of the displacement is within the range from 1 to 100 μm, giving rise to fretting damage.

Fretting wear involves various wear mechanisms such as adhesion, abrasion, oxidation and fatigue. At least two mechanisms occur simultaneously depending on operation parameters. Fretting wear produces oxidized wear debris and the wear coefficient increases rapidly with increasing amplitude. One of the main characteristics of reciprocating wear is that the wear coefficients are approximately constant for constant sliding distance and do not depend on sliding amplitude. The amplitude value for the transition between fretting and reciprocating is not easily defined as it depends on each set of experimental conditions and materials (Chen, 2001).

2.2.3 Wear mechanisms

Plastic deformation dominates when metals slide at low speed. The general requirement for plastic deformation is that the mechanical stresses generated by the adhesive, loading and frictional forces exceed the yield stress of one or both of the sliding materials. Plastic deformation leads to different surface damage forms:

- *Adhesion.* Adhesive forces combined with plastic flow pull out wear particles from the tips of the asperities.
- *Delamination.* Plastic flow nucleates and promotes the growth of subsurface cracks that propagate parallel to the surface, before extending out to the free surface to form platelet-like wear particles.
- *Surface cracks.* For brittle materials, the high tensile stresses generated by the traction within the contact region lead to surface fractures and cracks.
- *Fatigue.* Cyclic variations of stresses from repeated sliding or rolling contact results in fatigue failure in the near-surface region.
- *Mechanical milling and nanostructuring.* A mechanically mixed layer is formed and the composition of wear debris is a mixture of constituents from the two sliding surfaces. For unlubricated ductile materials, the grains near the surface and the wear debris can be refined down to the nanometer level.

- *Seizure*. Under extremely high contact pressures, the real area of contact approaches the nominal area leading to seizure of the moving parts followed by severe wear when they move again.

Adhesive wear

The concept of adhesive wear is based on the notion that adhesion occurs between asperities when they touch, followed by plastic shearing of the tips of the softer asperities which then adhere to the opposing surface, and finally separate as wear debris particles (Fig. 2.8). Adhesive wear can be described by the Archard model. A high surface energy should result in a high adhesive wear rate, as adhesive forces are higher for high surface energy. Since surface energy depends on the chemical composition of the surface, wear rates from this adhesive mechanism should be very sensitive to the presence of contamination layers or lubricant film.

Milling of materials is believed to be one of the causes of tribological surface film formation in sliding contacts. Besides the transfer and loss of material (Rigney, 2000), adhesive wear usually involves the following processes: adsorption of surface layers, i.e. moisture; sub-surface plastic deformation; microstructural and phase transformations; and surface film formation.

Figure 2.8 **Model for adhesion, transference of material and plastic deformation of wear debris**

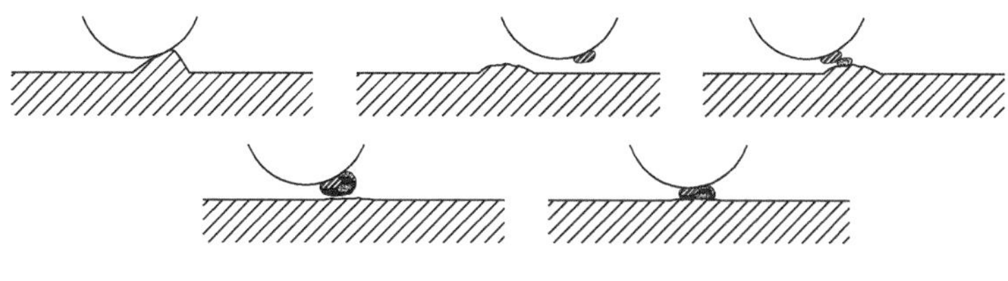

After the process of adhesion, transfer and plastic deformation, wear debris particles produced by adhesive wear show a flat rounded morphology (Williams, 2005), as can be observed in Fig. 2.9. Wear scars on the sliding surface show plastically deformed material on the edges of the track (Fig. 2.10 and Fig. 2.11).

Figure 2.10 shows a three-dimensional surface topography profile of a wear scar showing accumulation of plastically

Figure 2.9 SEM micrograph of an adhesive wear debris particle showing the flat rounded morphology

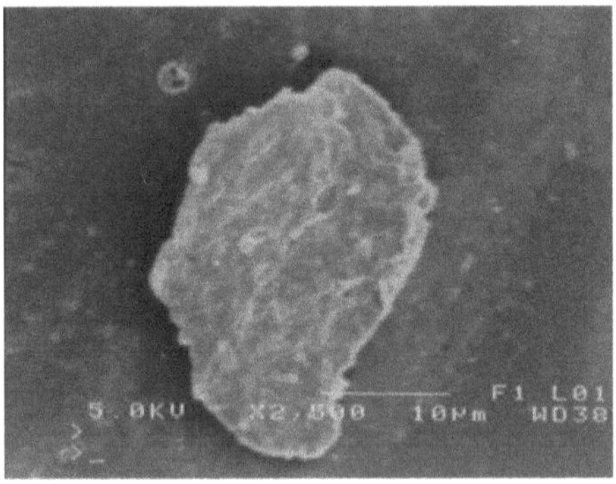

Figure 2.10 Three-dimensional surface topography profile of a dry wear scar

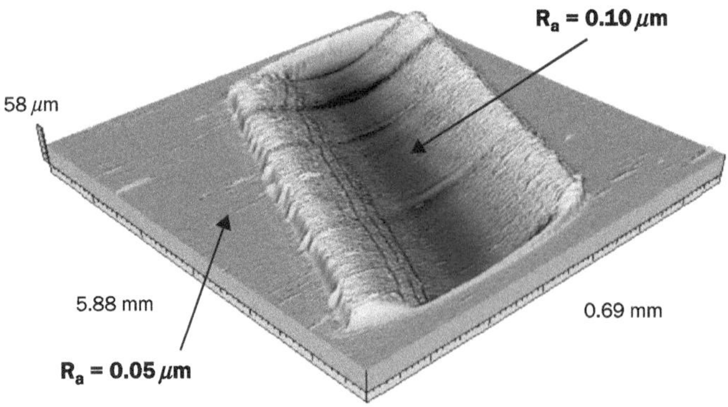

Figure 2.11 Line-scan of the cross-section of a wear scar

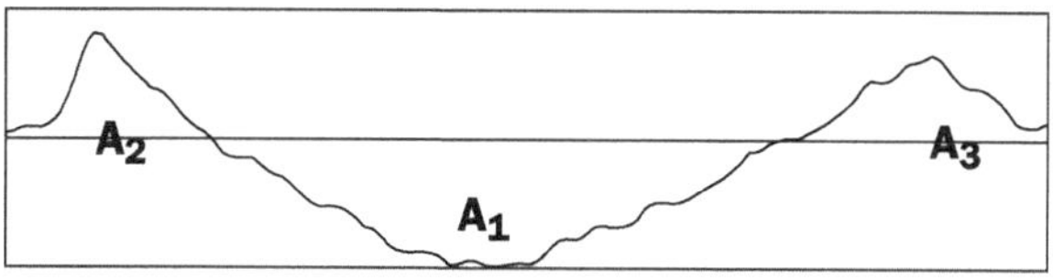

deformed material at the edges and the initiation of cracks perpendicular to the sliding direction. This wear mechanism produces an increase of the average roughness (R_a) inside the wear track with respect to the initial roughness outside the track.

The line-scan of the cross section of the wear track shown in Fig. 2.11 allows to determine the loss of material by measuring the area of material removed from the track surface (A_1) and the plastically deformed material ($A_2 + A_3$). The final amount of material lost by wear (A_T) will be

$$A_T = [A_1 - (A_2 + A_3)] \quad [2.18]$$

For metals sliding in air, adhesive wear can be reduced by the presence of a monolayer of boundary lubricant which not only reduces the surface energy but also resists displacement by the contacting asperities, thus preventing the direct contact between the metals.

Wear maps

Lim and Ashby (1987) developed a wear map for steel sliding against steel under a pin-on-disk configuration (Fig. 2.12). The map establishes the pressure-velocity regions where different wear mechanisms prevail and under relatively low pressures and velocities, material removal by *delamination* takes place. *Mild oxidational* wear occurs when oxidation due to frictional heating is relatively low, but under *severe*

Figure 2.12 **Load-velocity wear map for steel–steel under pin-on-disc configuration**

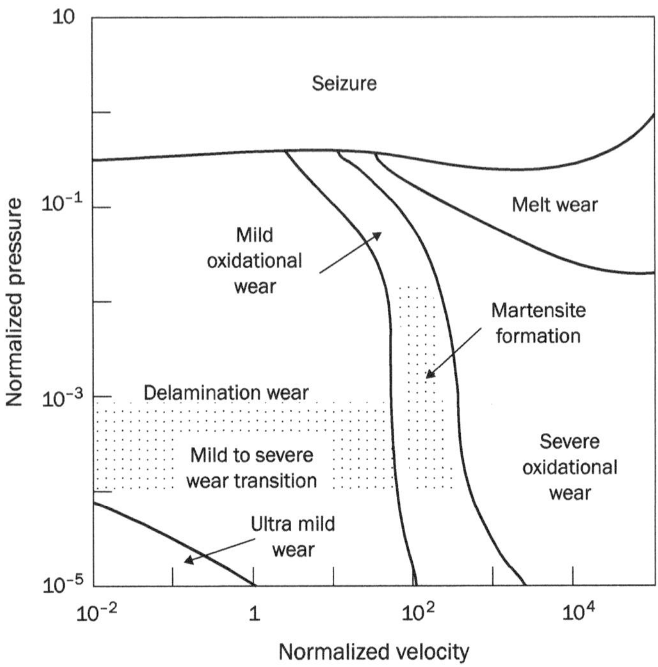

Adapted from Lim and Ashby (1987).

oxidational wear, wear particles are removed exclusively from the oxide layers. *Melt wear* occurs when flash temperatures reach the melting point. This map for steel was the precedent for other wear maps such as that developed by Adachi et al. (1997) for ceramic materials (Fig. 2.13).

In the case of ceramics, a quantitative model for the transition from mild to severe wear was developed from experimental data of ceramics sliding against themselves under the pin-on-disc configuration.

Abrasive wear

In abrasive wear, the rubbing of hard particles or hard asperities against a surface removes or displaces the material

Figure 2.13 Wear map for Al_2O_3

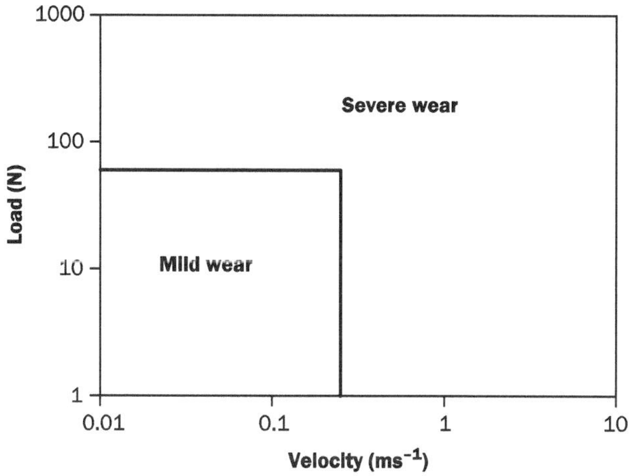

Adapted from Adachi et al., 1997

from that surface (Hokkirigawa and Kato, 1988). For abrasive wear to occur these particles, or one of the contacting surfaces, needs to be considerably harder than the surface being abraded (at least 1.3 times harder). The abrasive wear process leads to a characteristic surface topography of long grooves running in the sliding direction. The two general types of abrasive wear are:

- *Two-body abrasion*: Wear is caused by the hard protrusions of one surface on the other. Cutting tools produce two-body abrasive wear (Fig. 2.14).
- *Three-body abrasion*: Wear is caused by hard particles that slide between two sliding surfaces. This includes abrasion by wear debris generated by other wear mechanisms.

A simple way to model the abrasive wear (Siniawski et al., 2006) of an asperity in a two-body process or the corner of a grit particle in a three-body process is to consider a cone of a hard material gouging a groove into a softer material

Figure 2.14 Abrasive wear mechanisms by a sharp indenter

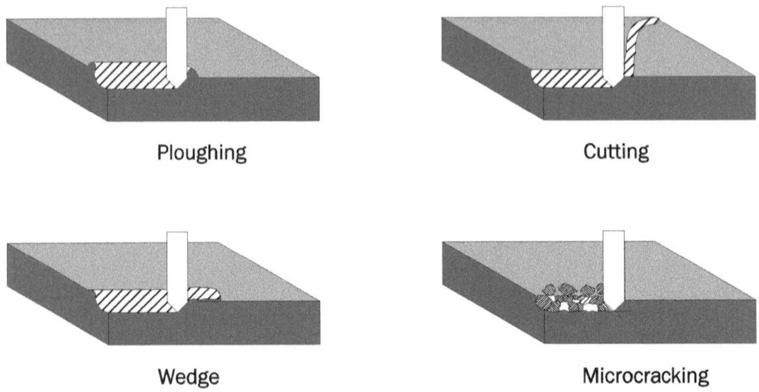

(Fig. 2.15). If the depth of the groove is x and the slope of the side of the cone is tan θ (Fig. 2.15), the volume V of material displaced when the asperity slides a distance d is given by

$$V = dx^2 \tan\varphi = dx^2 \cot\theta \qquad [2.19]$$

where 2φ is the cone angle of the indenter.

Figure 2.15 Abrasive wear by a cone-shaped asperity

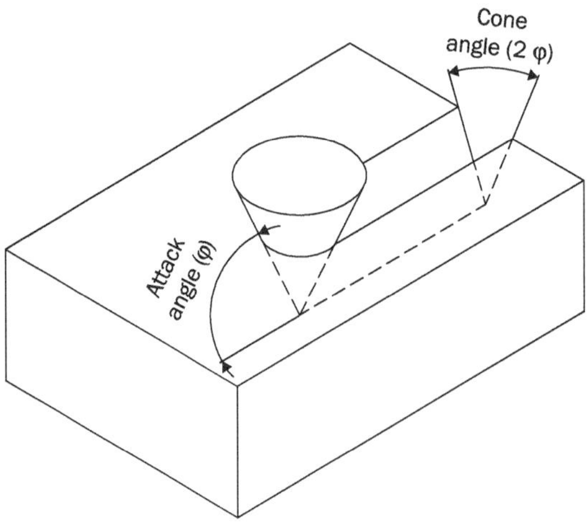

Since only the front half of the asperity applies a load W over an area $\pi a^2/2$, then

$$W = {}^1\!/\!_2 H\pi a^2 = {}^1\!/\!_2\, H\, \pi h^2 \cot^2\theta \qquad [2.20]$$

where H is the hardness of the softer material. From the above equations, the wear rate for abrasive wear (Q_a) will be

$$Q_a = V/d = 2\tan\theta W/\pi H \qquad [2.21]$$

and the coefficient of abrasive wear (K_a) would be $2/\pi(\tan\theta)$.

In contrast with the wear track profile seen in Fig. 2.11, abrasive wear produces parallel grooves inside the wear track (Fig. 2.16) without plastic deformation, and the morphology of wear debris particles (Williams, 2005) produced by abrasion (Fig. 2.17) is similar to that of the chips removed by machining operations. In most cases, however, a combination of adhesive and abrasive mechanisms is found due to the abrasion produced by the wear debris trapped in the interface between the sliding solids.

Scratch tests were developed to measure the resistance to abrasive wear of coatings. In the case of viscoelastic materials such as polymers (Brostow et al., 2002; Dasari et al., 2009),

Figure 2.16 Abrasive wear scar profile obtained by contact profilometry, showing abrasion grooves inside the scar and the absence of plastic deformation at the edges

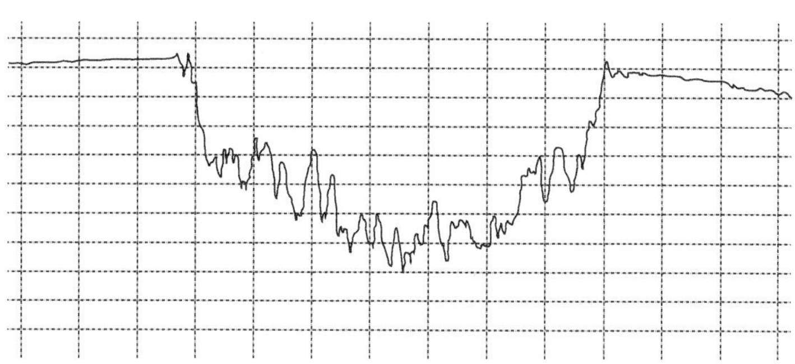

Figure 2.17 SEM micrograph of an abrasive wear debris particle produced by a cutting mechanism

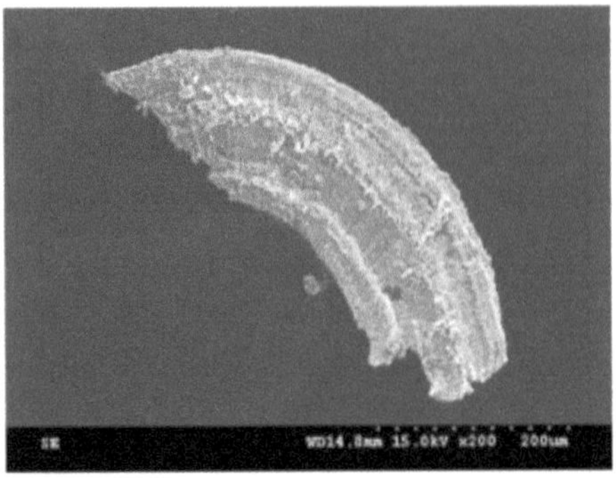

the resistance to abrasive wear is measured by a scratch test according to the scheme shown in Fig. 2.18. A hard indenter with a diamond tip registers the surface topology of the soft surface (line 1, Fig. 2.18), then the load F_n is applied and the penetration depth R_p is registered.

Viscoelastic materials recover or heal after the scratch, so the measurement of the residual depth R_h allows the determination of the extension of the recovery or healing of the material.

Figure 2.18 Scratch test configuration for viscoelastic materials

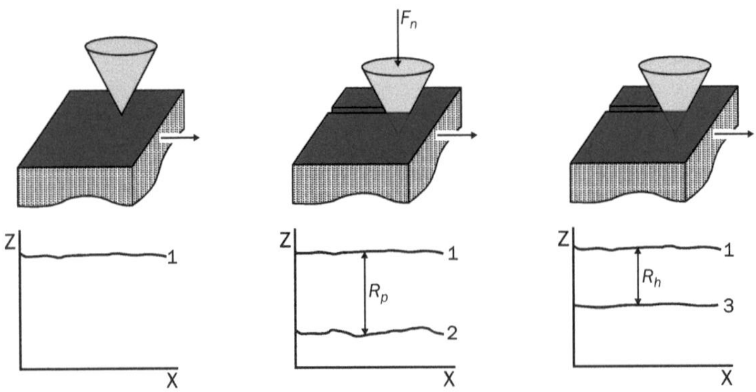

Erosive wear

Erosive wear takes place when hard solid particles impact on a surface (Tilly, 1977; Hutchings, 1992; Charles et al., 1997) and material loss due to erosion E is measured as the ratio of the mass of removed material with respect to the mass of erosive particles. Erosive wear is due to the impact of hard particles on surfaces. The erosion processes and erosive wear depend on the impact angle, impact velocity and relative hardness (Divakar et al., 2005) of the materials and incident particles:

$$E = cv^n f(\alpha) \quad [2.22]$$

where v is the particle velocity, α is the impact angle with respect to the material surface and the coefficient n varies from ductile to brittle materials.

Figure 2.19 illustrates the variation of the erosion rate with the impact angle for ductile and brittle materials. While ductile materials present the maximum erosion rate for an

Figure 2.19 Variation of erosive wear with impact angle for ductile and brittle materials

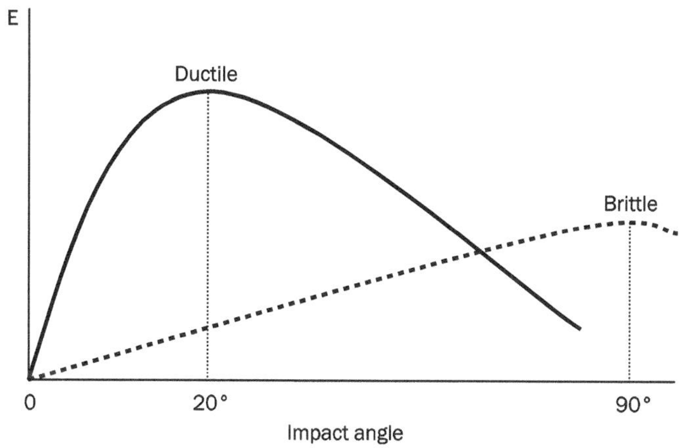

impact angle around 20°, and show a good erosion resistance for impacts normal to the surface, brittle materials show the more severe damage under normal impacts and have very good resistance at low angles. Ductile materials also present extensive plastic flow of the surface around the points of impact, while the impact of hard particles on brittle materials produces the propagation of cracks which lead to material removal from the surface.

Tribocorrosion

Protective films and passivating oxide layers can be removed by wear and the loss of material produced by wear is often associated with corrosion processes.

Materials and components failure by the synergistic combination of erosion processes and corrosive environments have been classified (*ASM Handbook*, vol. 18, 1992) as erosion-enhanced corrosion (EEC), when the damaged region is confined within the oxide scale, and corrosion-affected erosion (CAE), in which the damaged zone includes both the oxide scale and the metal (substrate).

Corrosion environments are varied and complex and require further deep treatment. Under dry (high temperature conditions), at least four regimes are necessary to describe tribocorrosion processes (Stack, 2002): pure erosion-dominated and pure corrosion-dominates regimed and two intermediate cases in which erosion prevails over corrosion or corrosion is more dominant than erosion.

2.3 References

Adachi, K., Kato, K. and Chen, N. (1997), 'Wear maps of ceramics', *Wear* 203–4: 291–301.

Anderson, S., Soderberg, A. and Björklund, S. (2007), 'Friction models for sliding dry, boundary and mixed lubricated contacts', *Tribol. Int.* **40**: 580–7.

Archard, J. F. (1953), 'Contact and rubbing of flat surfaces', *J. Appl. Phys.* **24**: 981–8.

ASM Handbook, vol. 18 (1992), 'Friction, lubrication and wear technology'. ASM.

Bayer, R. G. (2002), *Wear Analysis for Engineers*, HNB Publishing, New York.

Berman, A. D., Ducker, W. A. and Israelachvili, J. N. (1996), 'Origin and characterization of different stick-slip friction mechanisms', *Langmuir* **12**: 4559–63.

Blau, P. J. (2009a), *Friction Science and Technology: From Concepts to Applications*, 2nd edn, CRC Press, Boca Raton.

Blau, P. J. (2009b), 'Embedding wear models into friction models', *Tribol. Lett.* **34**: 75–9.

Bogdanovich, P. N. and Tkachuk, D. V. (2009), 'Thermal and thermomechanical phenomena in sliding contact', *J. Friction and Wear* **30**: 153–63.

Bowden, F. P. and Tabor, D. (1986), *Friction and Lubrication of Solids*, Oxford University Press.

Brostow, W., Bujard, B., Cassidy P. E., Hagg, H. E. and Montemartini, E. (2002), 'Effects of fluoropolymer addition to an epoxy on scratch depth and recovery', *Mat. Res. Innovat.* **6**: 7–12.

Bushan, B. (ed.) (2001), *Modern Tribology Handbook*, CRC Press, Boca Raton.

Charles, J. A., Crane, F. A. A. and Furness, J. A. G. (1997), *Selection and Use of Engineering Materials*, 3rd edn, Butterworth-Heinemann.

Chatelet, E., Michon, G., Manin, L. and Jacquet, G. (2008), 'Stick/slip phenomena in dynamics: choice of contact model. Numerical predictions and experiments', *Mech. Mach. Theory* **43**: 1211–24.

Chen, G. X. and Zhou, Z. R. (2001), 'Study on transition between fretting and reciprocating sliding wear', *Wear* **250**: 665–72.

Czichos, H. (1978), '*Tribology – A Systems Approach*', Elsevier.

Dasari, A., Yu, Z. Z. and Mai, Y. W. (2009), 'Fundamental aspects and recent progress on wear/scratch damage in polymer nanocomposites', *Mater. Sci. Eng. R-Reports* **63**: 31–80.

Divakar, M., Agarwal, V. K. and Singh, S. N. (2005), 'Effect of the material surface hardness on the erosion of AISI 316', *Wear* **259**: 110–17.

Hokkirigawa, K. and Kato, K. (1988), 'An experimental and theoretical investigation of ploughing, cutting and wedge formation during abrasive wear', *Tribol. Int.* **21**: 51–7.

Hsu, S. M. (1996), 'Fundamental mechanisms of friction and lubrication of materials', *Langmuir* **12**: 4482.

Hutchings, I. M. (1992), *Tribology: Friction and Wear of Engineering Materials*, Edward Arnold, London.

Kato, K. (2000), 'Wear in relation to friction – a review', *Wear* **241**: 151–7.

Kim, H. J. and Kim, D. E. (2009), 'Nano-scale friction: A Review', *Int. J. Precision Eng. Manufacturing* **10**: 141–51.

Lim, S. C. and Ashby, M. F. (1987), 'Overview no.55. Wear-mechanism maps', *Acta Metall.* **35**: 1–24.

Lovell, M. R. and Deng, Z. (2000), 'Experimental characterization of sliding friction: crossing from deformation to plowing contact', *J. Tribol. Trans. ASME* **122**: 856–63.

Mate, C. M. (2008), *Tribology on the Small Scale*, Oxford University Press.

Meng, H. C. and Ludema, K. C. (1995), 'Wear models and predictive equations: their form and concept', *Wear* **181**: 443–57.

Mo, Y., Turner, K. T. and Szlufarska, I. (2009), 'Friction laws at the nanoscale', *Nature* **457**: 1116–19.

Perry, S. S. and Tysoe, W. T. (2005), 'Frontiers of fundamental tribological research'. *Tribol. Lett.* **19**, 151–161.

Rapoport, L. (2009), 'Steady friction state and contact models of asperity interaction', *Wear* **267**: 1305–10.

Rice, S. L. and Moslehy, F. A. (1997), 'Modelling friction and wear phenomena', *Wear* **206**: 136–46.

Rigney, D. A. (2000), 'Transfer, mixing and associated chemical and mechanical processes during the sliding of ductile materials', *Wear* **245**: 1–9.

Röder, J., Bishop, A. R., Holian, B. L., Hammerberg, J. E. and Mikulla, R. P. (2000), 'Dry friction: modelling and energy flow', *Physica D* **142**: 306–16.

Santner, E., Klaffke, D., Meine, K., Polaczyk, Ch. and Spaltmann, D. (2005), 'Effects of friction on topography and viceversa', *Wear* **261**: 101–6.

Siniawski, M. T., Harris, S. J. and Wang, Q. (2006), 'A universal wear law for abrasion', *Wear* **262**: 883–8.

Stack, M. M. (2002), 'Mapping tribo-corrosion processes in dry and in aqueous conditions: some new directions for the new millenium', *Tribol. Int.* **35**: 681–9.

Tilly, G. P. (1977), 'Erosion caused by impact of solid particles', in *Treatise on Material Science and Technology*, vol. 13. Academic Press.

Williams, J. A. (1996), *Engineering Tribology*, Oxford University Press.

Williams, J. A. (2005), 'Wear and wear particles – some fundamentals', *Tribol. Int.* **38**: 863–70.

Yoshizawa, H., McGuiggan, P. and Israelachvili, J. (2003), 'Identification of a second dynamic state during stick-slip motion', *Science* **259**: 1305–8.

3

Lubrication and roughness

L. Burstein, Technion–IIT, Haifa, Israel

Abstract: This chapter introduces the basics of hydrodynamic lubrication theory and highlights its application to some machine parts with un-roughened (slider and cylindrical journal bearings) and roughened surfaces (parallel parts in relative motion). It first presents a theoretical analysis and results for two-dimensional sinusoidal and triangular wave roughened surfaces.

Keywords: hydrodynamic lubrication, slider and cylindrical bearings, sinusoidal roughness, triangular roughness, surfaces.

3.1 Introduction

From long ago to the present, lubrication has been an effective practical means for two simultaneous purposes: reduced friction losses and a longer service life of rubbing machine parts. To ensure optimal conditions for the machine parts, the lubricant's behaviour between the rubbing surfaces should be understood. Petrov (1883), Tower (1885), and Reynolds (1886) laid the foundations of lubrication theory, which still remains the main tool for calculation and design of rubbing machine elements. In this chapter, traditional and new problems in hydrodynamic lubrication theory are highlighted. In the first half the basics of lubrication and some 1D applications to slider bearings are described, while the second half presents applications to 2D lubrication of surfaces with sinusoidal and triangular roughness.

The following are basic terms used here and hereinafter.

Cavitation – an undesirable effect when the hydrodynamic pressure reaches, or drops below, its vapour pressure level; accompanied by the formation of vapour bubbles in a lubricating fluid; characterized by formation of cavities on the surface of a metallic hydraulic component.

Clearance – the minimal distance between opposite surface planes passing through the higher asperity peaks in the gap.

Control cell – a fragment of the surface-lube-surface system used as basis for study, calculations or comparisons.

Gap – the passage between the surfaces, filled by the lubricant.

Hydrodynamic pressure – the pressure in the lubricant when at least one of the surfaces is moving.

Lubrication – a process used to separate the opposite surfaces and reduce friction between machine parts, using a substance called 'lubricant' or 'lube' (e.g. oil).

Reference or relative parameters – the dimensional constant values used in constructing their dimensionless counterparts.

Roughness – the small amplitude deviations of a real surface; quantified by asperity height and width.

Roughness/asperity height – the vertical difference between the asperity peak and the real surface.

Wave – a regular disturbance of the surface, used here for description of its texture; quantified by its amplitude and wavelength.

Wave number – the total waves counted along one of the surface directions.

Wave ratio – a ratio of the asperity numbers of the two opposite surfaces (inter-surface wave ratio) or of two orthogonal directions of a surface (inner-surface wave ratio).

Wavelength/asperity width/asperity step – the spacing of adjoining peaks or hollows, or, in general, the spacing between pairs of points defining a period; sometimes abbreviated to 'wave', thus 'two waves' means the sum of two wavelengths.

3.2 Lubricants

The function of the lubricant is to keep the twin components of a part (in engines, bearings, reducers, etc.) separated. Lubricants are also used to reduce friction in machine-tool work (cutting, milling, grinding, etc.). A liquid lubricant contains typically 90 per cent oil and up to 10 per cent additives serving for reduced friction and wear, higher viscosity and resistance to corrosion, and other improved indices of the machine's operational functions.

Lubricants, according to their physical state, may be solid (e.g. graphite, molybdenum disulfide, hexagonal boron nitride, PTFE – polytetrafluoroethylene), liquid (automotive and other machine oils), and gaseous (carbon dioxide, nitrogen, inert gases). There are also intermediate types – semisolid/semiliquid (grease, powders, etc.). Lubricants are sometimes also classified according to their area of application – motor and transmission oils, foodstuffs, ointments, and plastics – and the first-named category, used in internal combustion engines, is the largest. Still another form of classification is the type of fluid base – lanolin, water, mineral oils, natural and synthetic oils.

Finally, there are the different national standards where the oils are designated by their viscosity. Thus, the American Society of Automotive Engineers (SAE) has eleven viscosity grades (SAE J300 specification), of which six are intended for winter use (SAE 0W, SAE 5W, SAE 10W, SAE 15W, SAE 20W, or SAE 25W) and five for summer use (SAE 20, SAE 30, SAE 40, SAE 50, and SAE 60). These numbers are a measure of the viscosity at a specific temperature. For example, the minimal kinematic viscosity of SAE 20 oil at 100 °C is 5.6 cSt (centistokes) and of SAE 40 is 12.5 cSt.

The ISO VG (International Organization for Standardization, viscosity grade) classification is appended to the above system.

There are also lube specifications on the basis of their service purpose. The American Petroleum Institute (API) currently has following classes for diesel engines: CJ-4, CI-4, CH-4, CG-4, CF-2, and CF ('C' standing for 'commercial'). More detailed information about these and other oil data and classifications are available in specialized literature (see, for example, Booser, 1988). Knowledge of lubricant classification, labelling and properties allows to select the effective lube for a specific machine ensuring lower friction and a longer service life.

3.3 Regimes of lubrication

With regard to lubricant film thickness between surfaces in relative motion, four regimes can be recognized: hydrodynamic, boundary, mixed, and elastohydrodynamic. The parameter usually used for characterizing these regimes is the ratio of the minimal film thickness h to the asperity height R_a. A commonly accepted illustration of the regimes is the Stribeck curve (Fig. 3.1) which represents the dependence of the friction coefficient on the so-called bearing ratio $\eta u/\bar{p}$ where η is the kinematic viscosity, u the velocity of the moving part, and $\bar{p}$ the mean pressure in the fluid film).

3.3.1 Hydrodynamic lubrication regime

The hydrodynamic lubrication regime, $h >> R_a$, occurs when the lubricant completely separates the surfaces (see Fig. 3.1); it is mostly associated with film thicknesses near or more than 1 μm.

Friction losses under hydrodynamic lubrication are very small – less than in the other lubrication regimes. The hydrodynamic pressure generated in the lubricant film duc to the relative motion and inner film friction is too low to cause surface deformation. The flow of the lubricant film is laminar, but at thicknesses above 20 μm it becomes turbulent and that leads to undesirable friction losses.

Figure 3.1 **Stribeck curve and lubrication regimes**

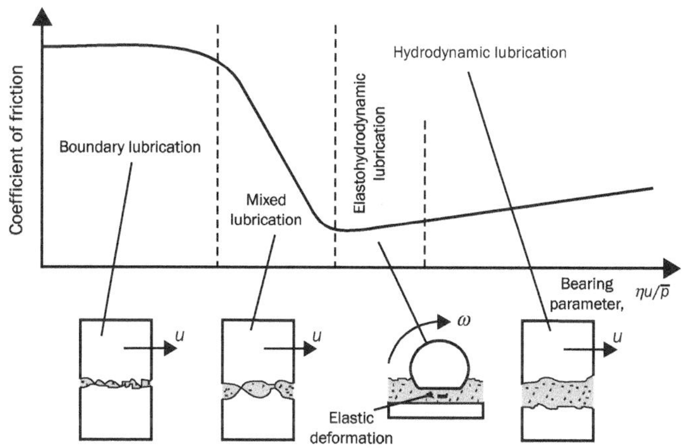

3.3.2 Boundary lubrication regime

The boundary regime, $h < R_a$, occurs when the fluid film is discontinuous and permits direct contact between high points (known as asperities) of the opposite surfaces (Fig. 3.1). It is characterized by film thicknesses less than 70 nm and higher friction losses than under the other regimes. Examples when this may occur are during equipment startup or shutdown, when the bearing may operate in boundary rather than in unbroken fluid film conditions, or in toothed gear contact, or in reciprocating motion (e.g. car valve on valve seat). In this mode the film has less carrying capacity than with contacted asperities.

3.3.3 Mixed lubrication regime

The mixed regime, $h \sim R_a$, is transitional between the boundary and hydrodynamic, when the fully-lubricated (separated) and contacted (unseparated) surface areas equally influence the friction and the film parameters. Film thicknesses range from above 70 nm to 1 μm (Fig. 3.1). Friction losses vary over a wide range according to the two limiting regimes involved.

3.3.4 Elastohydrodynamic lubrication regime

The elastohydrodynamic (EHD) regime is a particular case of the hydrodynamic with high hydrodynamic pressures, sufficient to impact one or two of the opposite surfaces. The latter are separated, but there is some interaction between the asperities, resulting in elastic deformation and enlargement of the contactless area (Fig. 3.1), whereby the viscous resistance of the lubricant becomes capable of supporting the load. Film thicknesses usually range from 10 to 70 nm; the minimal thickness is often connected to the film's carrying capacity by the relation $h_{min} \sim w^{-0.073}$. The losses are the same as under hydrodynamic lubrication, as the surfaces are completely separated. The main assumption for the theoretical analysis is immediate elastic deformation of the contacting materials and immediate increase of the lubricant viscosity.

3.4 Reynolds' equation

The flow of the lubricant obeys the classical hydrodynamic theory with some simplifications specific to thin fluids. The surfaces, the fluid film thickness, and the fluid- and moving-surface velocities are shown schematically in Fig. 3.2. Here we proceed to derive an expression that describes the hydrodynamic pressure in lubrication, making use of the momentum-transport (Navier-Stokes) and continuity equation.

Figure 3.2 **Derivation scheme for Reynolds' equation. Surfaces, fluid film thickness, velocities and coordinates**

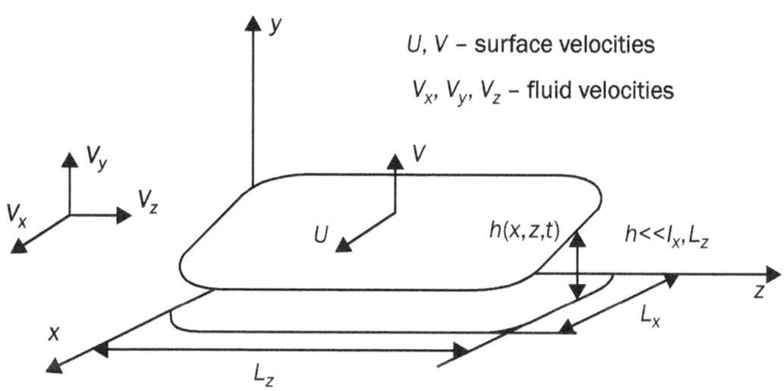

The first of these in vector form reads:

$$\frac{\partial \vec{v}}{\partial t} = f - \frac{1}{\rho}\overrightarrow{\text{grad}\, p} + \nu \Delta \vec{v} \tag{3.1}$$

where p is hydrodynamic pressure, v is flow velocity, t is time, f are forces acting on the fluid, ρ is fluid density, ν is kinematic viscosity, Δ is Lapltacian $\frac{\partial^2}{\partial x^2}+\frac{\partial^2}{\partial y^2}+\frac{\partial^2}{\partial z^2}$, grad is gradient $\frac{\partial}{\partial x}i+\frac{\partial}{\partial y}j+\frac{\partial}{\partial z}k$, and i, j, k are unit vectors of the x–, y–, and z directions respectively.

For an incompressible, isoviscous Newtonian fluid in laminar flow, its components can be described in full format as:

$$\begin{aligned}
\rho\left\{\frac{\partial v_x}{\partial t}+v_x\frac{\partial v_x}{\partial x}+v_y\frac{\partial v_x}{\partial y}+v_z\frac{\partial v_x}{\partial z}\right\} &= -\frac{\partial p}{\partial x}+\mu\Delta v_x \\
\rho\left\{\frac{\partial v_y}{\partial t}+v_y\frac{\partial v_y}{\partial x}+v_y\frac{\partial v_y}{\partial y}+v_z\frac{\partial v_y}{\partial z}\right\} &= -\frac{\partial p}{\partial y}+\mu\Delta v_y \\
\rho\left\{\frac{\partial v_z}{\partial t}+v_z\frac{\partial v_z}{\partial x}+v_z\frac{\partial v_z}{\partial y}+v_z\frac{\partial v_z}{\partial z}\right\} &= -\frac{\partial p}{\partial z}+\mu\Delta v_z
\end{aligned} \tag{3.2}$$

where μ is the dynamic viscosity.

As regards the continuity equation, for laminar inertialess flow it reads:

$$\frac{\partial(\rho v_x)}{\partial x}+\frac{\partial(\rho v_y)}{\partial y}+\frac{(\rho v_z)}{\partial z}=0 \tag{3.3}$$

Because of the smallness of the gap-to-length ratio $(h/l) << 1$ (the ratio of film thickness to its length) the flow terms of order $(h/l)^2$ and higher can be disregarded in the eqs. [3.2], presented previously in dimensionless form. Thus, eqs. [3.2], rewritten for inertialess $\frac{\partial v_x}{\partial t}=\frac{\partial v_y}{\partial t}=\frac{\partial v_z}{\partial t}=0$ flow, reduce to:

$$\frac{\partial p}{\partial x} = \mu \frac{\partial^2 v_x}{\partial y^2}$$
$$\frac{\partial p}{\partial z} = \mu \Delta \frac{\partial^2 v_y}{\partial y^2} \qquad [3.4]$$
$$\frac{\partial p}{\partial y} = 0$$

with the boundary conditions

$$v_x = 0;\ v_y = 0 \quad \text{at}\ y = 0$$
$$v_x = \mathrm{u};\ v_y = 0 \quad \text{at}\ y = \mathrm{h} \qquad [3.5]$$

Integrating the continuity equation with respect to y over the thickness limits 0 . . . h (the pressure and $\frac{\partial p}{\partial y}$ do not change in that dirction) and applying Leibnitz's integration formulae $\int_0^{h(\varsigma)} \frac{\partial g}{\partial \varsigma} \mathrm{dy} = \frac{\partial}{\partial \varsigma} \int_0^{h(\varsigma)} g\mathrm{dy} - g(\varsigma, h)\frac{\partial h}{\partial \varsigma}$, we obtain

$$\frac{\partial}{\partial x}\left(\int_0^h \rho v_x\, \mathrm{dy}\right) - \rho v_x \frac{\partial h}{\partial x} + \frac{\partial}{\partial z}\left(\int_0^h \rho v_z\, \mathrm{dy}\right)$$
$$- \rho v_z \frac{\partial h}{\partial z} + \rho v_{y=h} = 0 \qquad [3.6]$$

Here the velocity $\mathrm{v}_{y=h}$ is the total derivative, or in partial form

$$\mathrm{v}_{y=h} = \frac{\partial h}{\partial t} + u\frac{\partial h}{\partial x} \qquad [3.7]$$

Further, the integrals in eq. [3.6] represent the mass flow rates (lube consumption) in the x and z directions. Substituting the velocities (determined from eqs. [3.4]) and considering the boundary conditions [3.5], we can obtain

$$M_x = \int_0^h \rho v_x \, dy = \int_0^h \left\{ \frac{\rho}{2\mu} \frac{\partial p}{\partial x} \left[y^2 - yh \right] + \frac{y}{h} u_x \right\} dy$$

$$= \frac{\rho h^3}{12\mu} \frac{\partial p}{\partial x} + \frac{\rho h u}{2} \qquad [3.8]$$

$$M_z = \int_0^h \rho u_z \, dy = \int_0^h \frac{\rho}{2\mu} \frac{\partial p}{\partial z} \left[y^2 - yh \right] dy = -\frac{p h^3}{12\mu} \frac{\partial p}{\partial z}$$

Note now that $\frac{\partial(\rho v_x h)}{\partial x} + \frac{\partial(\rho v_z h)}{\partial z} = 0$, the law of mass conservation. Considering this in conjunction with eqs. [3.7] and [3.8], eq. [3.6] transforms into the elliptical differential equation obtained by Reynolds, named after him, and known as the central expression of hydrodynamic lubrication theory:

$$\frac{\partial}{\partial x}\left(h^3 \frac{\partial p}{\partial x} \right) + \frac{\partial}{\partial z}\left(h^3 \frac{\partial p}{\partial z} \right) = 6\mu u \frac{\partial h}{\partial x} - 12\mu \frac{\partial h}{\partial t} \qquad [3.9]$$

The pressures on the left-hand side generate the flow through their gradients; the right-hand terms represent the flow generated by a possible shift in the *y*-direction (squeeze) and by the moving- surface velocity u. The fluid velocities

$$\mathrm{v}_x = \frac{1}{2\mu} \frac{\partial p}{\partial x} \left(y^2 - yh \right) + \frac{y}{h} u$$

$$\mathrm{v}_z = \frac{1}{2\mu} \frac{\partial p}{\partial z} \left(y^2 - yh \right) \qquad [3.10]$$

reflect the superposition of two flow modes: the Poiseuille, due to imposed pressure gradients; and the Couette, a shear-driven effect due to motion of the upper surface. In the v_x curve the Poiseuille flow is associated with a parabolic velocity distribution across the film, and the Couette flow with a linear one.

Solution of Reynolds' equation yields the hydrodynamic pressure distribution which in turn, combined with eqs. [3.10], yields the film's load capacity

$$w = \int \int p\mathrm{d}x\mathrm{d}y \quad [3.11]$$

as well as the shear stress and the friction force at the lubricant-surface contact:

$$\tau = \mu \left(\frac{\partial u}{\partial y} \right)_{y=h}$$
$$f_f = \int \tau \mathrm{d}x \quad [3.12]$$

Solution of Reynolds' equation involves boundary conditions specific to each practical case. How this is done is shown in the subsequent sections, on the examples of slider and cylindrical journal bearings.

3.5 Applications of hydrodynamic lubrication theory

In view of its key role in the pressure distribution determination, Reynolds' equation is instrumental in designing surfaces with optimal geometries. Such is the case with bearings, mechanical seals, gears, jet engines (e.g. turbine blades), internal combustion engines (e.g. piston-cylinder and ring-cylinder couples, crank mechanisms), nanotechnology, and also in biomedicine (e.g. artificial joints, synovial fluid), and many other areas.

Below, a concise analysis of 1D fluid flow in slider bearings is presented. These parts appear in a variety of configurations, e.g. plain and step bearings, plain journal and partial arc bearings, axial groove, offset half, four-lobe, tilting pad and others. In all of them, only one member of the couple is in motion.

3.5.1 *Slider bearings*

Figure 3.3 shows the geometry of a 1D slider (thrust) bearing, with the coordinate system used for solving Reynolds' equation.

Figure 3.3 Slider geometry and coordinates

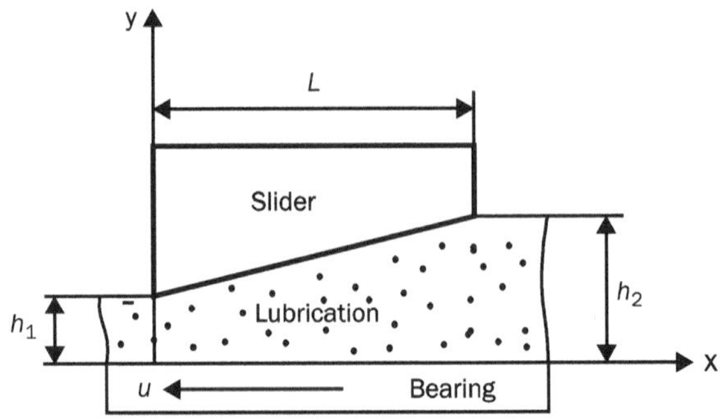

The film thickness (h) has a linear taper codirectional with the surface velocity u

$$h = h_1 + \left(\frac{h_2 - h_1}{B}\right)x \qquad [3.12]$$

Note that the exit film thickness h_2 is unknown and should be determined as part of the bearing design. The bearing taper (h_1-h_2) is also a design parameter (determined via analysis).

Reynolds' equation for this 1D case has the form

$$\frac{\partial}{\partial x}\left(h^3 \frac{\partial p}{\partial x}\right) = -6\mu u \frac{\partial h}{\partial x} \qquad [3.13]$$

Transformation of eqs. [3.12] and [3.13] with $X = x/L$, $H = h/h_1$, $P = ph_1^2/(6\mu uL)$, $k = (h_2 - h_1)/h_1$ leads to the dimensionless expressions

$$H = 1 + kX \qquad [3.14]$$

and

$$\frac{\partial}{\partial X}\left(H^3 \frac{\partial P}{\partial X}\right) = -\frac{\partial H}{\partial X} \qquad [3.15]$$

Substituting the derivative of the thickness and passing to those with respect to H according to the chain rule $\frac{\partial}{\partial X} = \frac{\partial}{\partial H}\frac{\partial H}{\partial X}$, we obtain

$$\frac{\partial}{\partial H}\left(H^3\frac{\partial P}{\partial H}\right)=-\frac{1}{k} \qquad [3.16]$$

And after double integration with respect to H

$$P=\frac{1}{kH}-\frac{C_1}{2H^2}+C_2 \qquad [3.17]$$

The constants of integration should be determined from the boundary conditions $P=0$ at both $X=0$ and $X=1$ (no hydrodynamic pressure outside the slider wedge). Applying this to eq. [3.15], we obtain $C_1=-2(1+k)/[k(2+k)]$ and $C_2=-1/[k(2+k)]$. Thus

$$P=\frac{1}{k}\left[\frac{1}{H}-\frac{1}{2+k}-\frac{1+k}{2+k}\frac{1}{H^2}\right] \qquad [3.18]$$

The hydrodynamic pressures for several p calculated by eq. [3.18] are presented in Fig. 3.4 for several slider inlet/outlet ratios $H_2/H_1=k+1$. Increase of the maximal pressure

Figure 3.4 Dimensionless pressure and inlet-to-outlet ratio

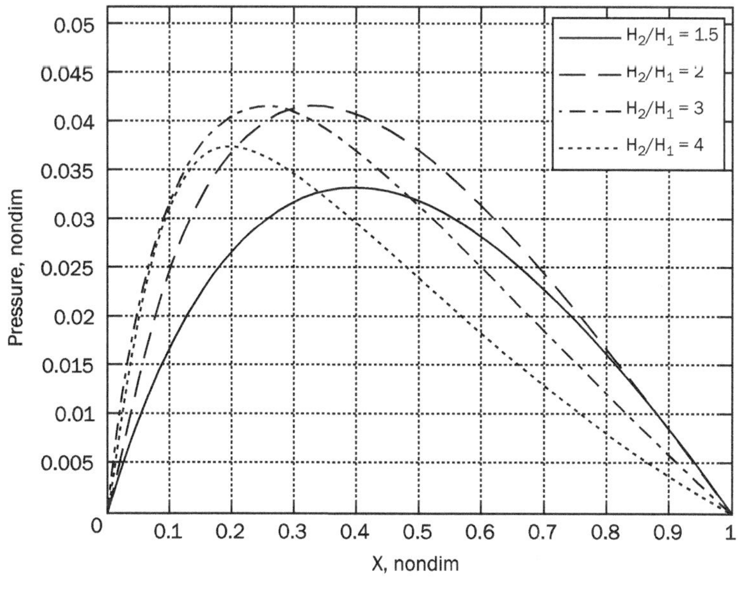

with the ratio H_2/H_1 occurs at small wedge angles (ratio less than 2.2), above which it drops – see curve for $H_2/H_1 = 4$.

3.5.2 Cylindrical journal bearings

Cylindrical journal bearings are common in rotating machinery such as generators, motors, compressors, turbines, etc. This bearing comprises of an inner rotating cylinder (journal) of radius R and an outer immobile cylinder (bearing) of radius $R_b > R$. The geometry and coordinate system are shown in Fig. 3.5.

In most fluid-film bearings with incompressible liquid lubricant, the radial clearance ratio $(R_b - R)/R_b$ is typically 0.001. When $x = R\theta$, $z = ZL$, $t = TR/u$, $h = Hc$, and $P = pc^2/(6\mu uR)$, Reynolds' equation [3.9] can be rewritten in dimensionless form as:

$$\frac{\partial}{\partial\theta}\left(H^3\frac{\partial P}{\partial\theta}\right) + \ell_b^2\frac{\partial}{\partial Z}\left(H^3\frac{\partial P}{\partial Z}\right) = \frac{\partial H}{\partial\theta} - 2\frac{\partial H}{\partial T} \qquad [3.19]$$

Figure 3.5 Cylindrical journal bearing geometry and coordinate system

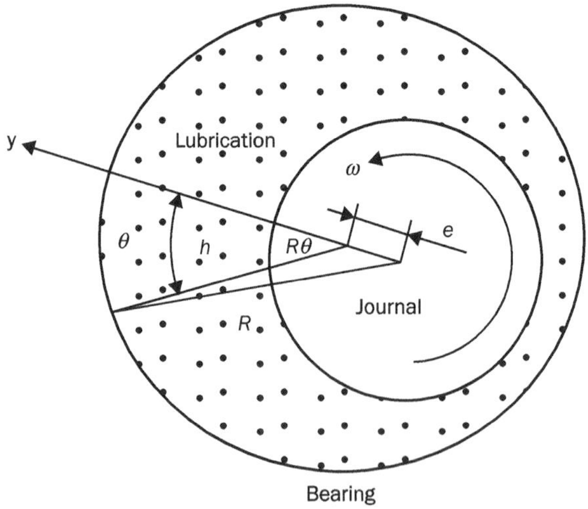

where $\ell_b = R/L$ is the bearing scale factor, L is the bearing length, and the velocity can be defined as $u = R\omega$.

The gap between journal and bearing can also be presented in dimensionless form, using the notations e for eccentricity of journal center, c for clearance ($c = R_b - R$), $H = h/c$, $\varepsilon = e/c$, and representing the cylindrical surfaces by planes, as follows:

$$H = 1 + \varepsilon \cos\theta \qquad [3.20]$$

This expression is justified usually at c/R_b ratios smaller than 0.1.

Assuming that the bearing parameters in the θ-direction (x,y-plane) are less than the bearing length; a $\ell_b < 1$ (long bearing, thus the pressure does not change along Z and $\partial P/\partial Z = 0$); and when the gap configuration does not change with time, $\partial H/\partial T = 0$, eq. [3.19] simplifies to

$$\frac{\partial}{\partial\theta}\left(H^3 \frac{\partial P}{\partial\theta}\right) = \frac{\partial H}{\partial\theta} \qquad [3.21]$$

The boundary conditions for this case are $P(\theta) = P(2\pi) = P_0$ and called Sommerfeld's conditions. After the first integration we have

$$\frac{\partial P}{\partial\theta} = \frac{1}{H^2} + \frac{C_1}{H^3} \qquad [3.22]$$

and further integration of both parts of eq. [3.22] yields

$$P = \int \frac{d\theta}{(1+\varepsilon\cos\theta)^2} + C_1 \int \frac{d\theta}{(1+\varepsilon\cos\theta)^3} + C_2 \qquad [3.23]$$

The integration constants are determined subject to the above boundary conditions, namely

$$C_1 = \frac{\int_0^{2\pi} \frac{d\theta}{(1+\varepsilon\cos\theta)^2}}{\int_0^{2\pi} \frac{d\theta}{(1+\varepsilon\cos\theta)^3}} = \frac{2(1-\varepsilon^2)}{2+\varepsilon^2} \qquad [3.24]$$

$$C_2 = P_0$$

The integrals in eqs. [3.23] and [3.24], known as the journal-bearing integrals, are obtainable for instance from Booker, 1965. Denoting the hydrodynamic pressure at $\theta=0$ by P_0 (pressure at the initial angle), we have the pressure distribution equation as

$$P = P_0 + \frac{\varepsilon(2+\varepsilon\cos\theta)\sin\theta}{(2+\varepsilon^2)(1+\varepsilon\cos\theta)^2} \quad [3.25]$$

This solution was first arrived at by Sommerfeld in 1904, and remains topical to this day.

The redundant hydrodynamic pressures, P-P_0, calculated by eq. [3.25], are presented in Fig. 3.6 for several values of the dimensionless eccentricity ε. The pressure in the solution has positive and negative branches with equal maximum and minimum. As can be seen, the pressure increases with the

Figure 3.6 **Dimensionless pressure distribution as function of angle at different eccentricities**

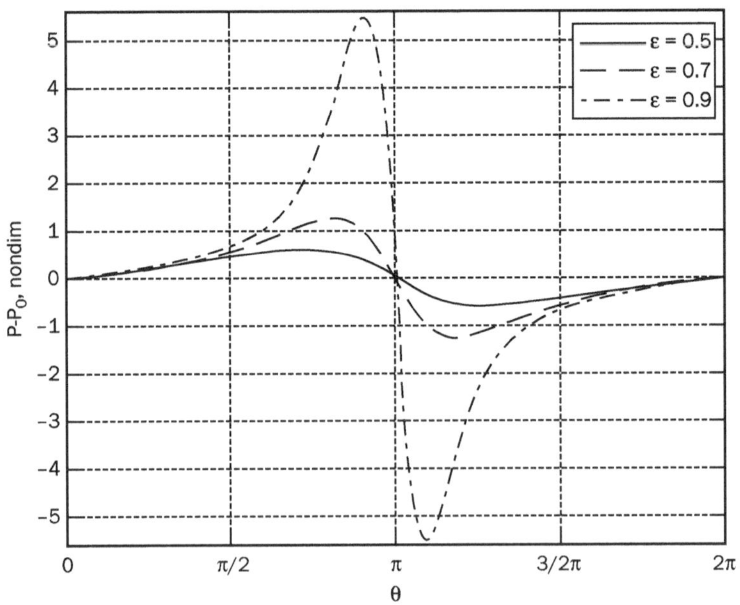

eccentricity; this fact should be borne in mind in the bearing design when choosing the operating eccentricity.

3.6 Hydrodynamic lubrication of roughened surfaces

The equations derived in the preceding sections are valid for ideally smooth surfaces, which is not the case in practice. Under the microscope we can see micro-deviations from the ideal form; a schematic view of the roughness traced by measurement stylus, together with some roughness parameters, is presented in Fig. 3.7. Here R_a is the mean asperity height and S_i and h_i are the local roughness step and height. There are also other roughness parameters specified in various national and international standards (e.g. ANSI, DIN, ISO). So, the R_a-values specified by ISO are related to the roughness grade number N ($N1$, $N2$, . . ., up to $N11$) and range from 0.025 up to 25 µm accordingly. The mean roughness step (interval between asperities), named

Figure 3.7 Real surface profile and some roughness parameters

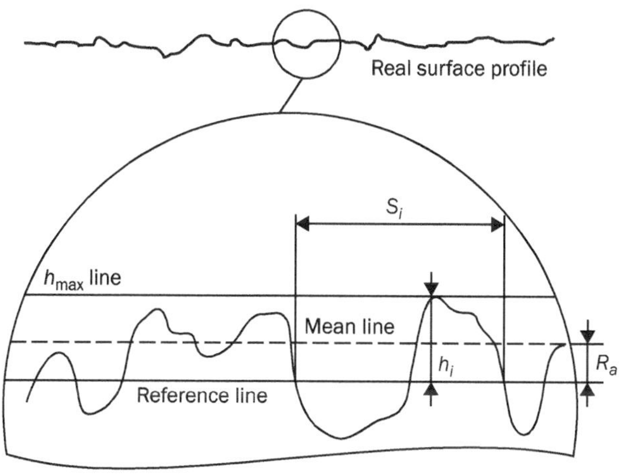

sometimes roughness wavelength, varies in the wide range 3.5–75 μm; both roughness parameters are connected with the type of surface finish, from lower values for fine machining to higher values for rough machining.

The size of the surface asperities often equals and even exceeds the film thickness (see Section 3.2), thus they naturally influence the hydrodynamic behaviour of the film. Knowing how the roughness parameters, asperity height R_a and step S influence the film pressure, the data obtained for smooth surfaces can be adjusted with a view to more reliable design of the machine parts.

Hydrodynamic lubrication of roughened surfaces has long been an object of study at fluctuating levels of intensity, see for example Dowson et al. (1978). With the advent of laser machining, interest in it has peaked again. A flow model was developed by Patir and Cheng (1978, 1979) using averaged roughness indices, and later gradually replaced by deterministic models based on detailed roughness geometry. This approach had already been attempted in early studies of microasperities and microcavities (Anno and Walowit, 1968, 1969), and more recently in that of regularly and irregularly covered porous surfaces (Lai, 1994; Burstein and Ingman, 1999, 2000; Arghir et al., 2003). Common to these studies is emphasis on the integrated surface behaviour, it being assumed that the maximal load support found for one microgroove is the same for all the others; the cumulative effect is not considered. In general, it is necessary to examine the entire profile or, in three-dimensional terms, the entire real surfaces separated by the lube.

Most recently in the last few decades, we are witnessing a new approach whereby the opposite surfaces are deterministically described by some idealized but unified expression, thus permitting in principle a theoretical solution (Labiau et al., 2008; Letallear et al., 2002; Olver and Dini, 2007). One common model for regular structure surfaces is the standardized version (e.g. ANSI/ASME Standard, 1985;

SRM 2073a, 1995) where the roughness is presented as a sinusoidal ripple surface with a given frequency (Tønder, 1999, 2004) obtainable experimentally, e.g. the main frequency of the spectral density function determined from the spectral diagram of the concrete surface.

In parallel, 1D and 2D transient lubrication was theoretically studied for surfaces with equal asperity sizes, and finally generalized in a model representing unequal roughnesses and unequal wavelengths along the axes – of which the equal-roughness version is a particular case (see Burstein, 2006, 2008, 2009). In these studies the opposite surfaces also had equal wave numbers along the identical axes. Under the generalized approach, sinusoidal roughness and the same number of waves were assumed on both surfaces. Lately non-sinusoidal 1D and 2D roughness models were developed (Burstein, 2010), in which it was possible to study the unequal case for both the asperity heights and wavelengths. Hereafter two roughness types are briefly presented: sinusoidal and triangular wave configurations.

3.6.1 Roughness parameters and surface roughness models

As was noted earlier, even perfectly finished machine parts have haphazardly distributed numerous micro or nano peaks and hollows reflected in a similar pattern in the gap width and correspondingly in the film thickness geometry, with the attendant difficulties in theoretical and computerized treatment. The main problem is that the randomly generated opposite surfaces (see Fig. 3.8, surface generated by Matlab) are in relative motion with the gap changing from moment to moment, so that Reynolds' equation cannot be solved analytically. Numerical 1D or 2D solutions are also highly problematic because of the large numbers of asperities on the real surfaces.

Figure 3.8 **Computer image of surface with randomly generated asperity heights and roughness step**

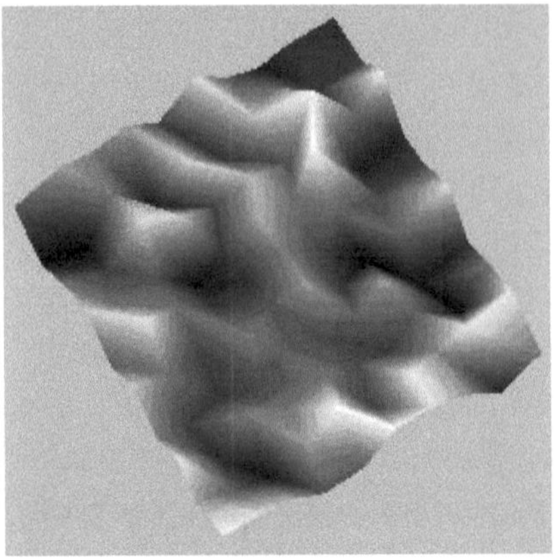

Note: Asperity heights 2μm, roughness step 2.5R_a, standard deviation 0.2 of the asperity height; five asperities in each dimension; azimuth –31, elevation 86

The problem becomes manageable if the profile is approximated by some regular configuration, mostly by a sinusoid or by a triangular wave function. A computer-generated view of such surfaces, modelled in Matlab, is presented in Fig. 3.9(a) and (b). In spite of the outward similarity of these surfaces, the triangular pattern is maybe more realistic than the sinusoidal – see Fig. 3.7. In any event, applying these models permits better evaluation of the surface indices, selecting those suitable for each specific case.

Figure 3.9 **Surface with sinusoidal (a) and triangular (b) roughness**

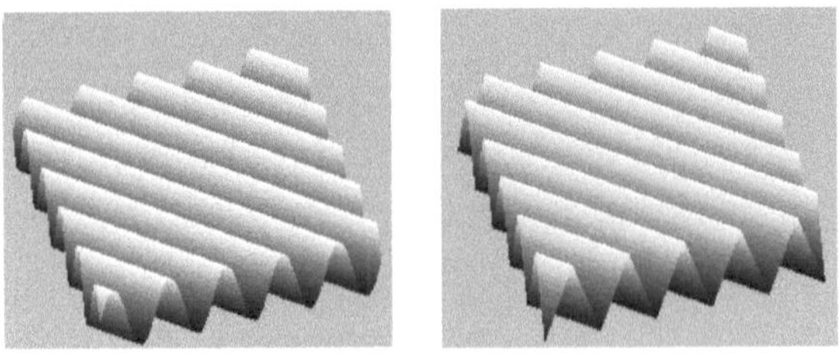

(a) (b)

Note: Asperity heights 2μm and roughness step 2.5R_a, five waves in each dimension; azimuth –17, elevation 66

3.6.2 Hydrodynamic lubrication for sinusoidal roughness

Reynolds' equation can be rewritten in dimensionless form as follows:

$$\frac{\partial}{\partial X}\left(H^3\frac{\partial p}{\partial X}\right)+\ell_S^2\frac{\partial}{\partial Z}\left(H^3\frac{\partial p}{\partial Z}\right)=\frac{\partial H}{\partial X}-2\frac{\partial H}{\partial T} \qquad [3.26]$$

where

$X=X/L_x,\ Z=z/L_z,\ T=ut/L_x,\ H=h/\left[\Delta+R_{a_1}+R_{a_2}\right],$

$P=p\left(\Delta+R_{a_1}+R_{a_2}\right)^2/6\mu L_x u, \ell_S=L_x/L_z,$ and subscripts 1 and 2 denote the stationary and moving surfaces respectively. The x-plane of the profile geometry and the 3D view of the surfaces are presented in Fig. 3.10.

With the surfaces in sinusoidal form, the dimensionless gap is given by

$$\begin{aligned}H=1+A_2\sin 2\pi k_{x_2}\left(X+\ell_2^{-1}Z-T+\Phi_2\right)\\ -A_1\sin 2\pi k_{x_1}\left(X+\ell_1^{-1}Z\right)\end{aligned} \qquad [3.27]$$

where A_1 and A_2 are the dimensionless asperity heights, $A_1=R_{a_1}/(\Delta+R_{a_1}+R_{a_2})$, $A_2=R_{a_2}/(\Delta+R_{a_1}+R_{a_2})$, and $\ell_1=k_{x_1}/k_{z_1}$ and $\ell_2=k_{x_2}/k_{z_2}$ are the wave ratios; k_{x_1}, k_{x_2}, k_{z_1}, k_{z_2} are the wave numbers along the X and Z axes for surfaces 1 and 2; $\Phi=\varphi/(2\pi L_x)$ – the phase angle of the moving surface. Here the roughnesses – R_{a_1}, R_{a_2} and the wave numbers k_{x_1}, k_{z_1}, k_{x_2}, k_{z_2} – differ between the surfaces; and the origin of coordinates is set such that the phase angle $\Phi_1=0$.

Assuming that the surfaces have the same width and length $L_x=L_z=L$ or $\ell_S=1$, and the same wave ratios $\ell_1=\ell_2=\ell$, we can concentrate on two cases – (a) the wave numbers in the X- and Z-directions being equal (scale factor $\ell=1$) and (b) the wave numbers unequal ($\ell_1\neq\ell_2$, scale factor $\ell\neq 1$).

Figure 3.10 Schematic of unequally roughened surfaces (a) and gap geometry in X,H plane (b) at wave number $k=5$, roughness height ratio $R_{a_1}/R_{a_2}=0.5$, and phase displacement $\Phi=1/(4k)$

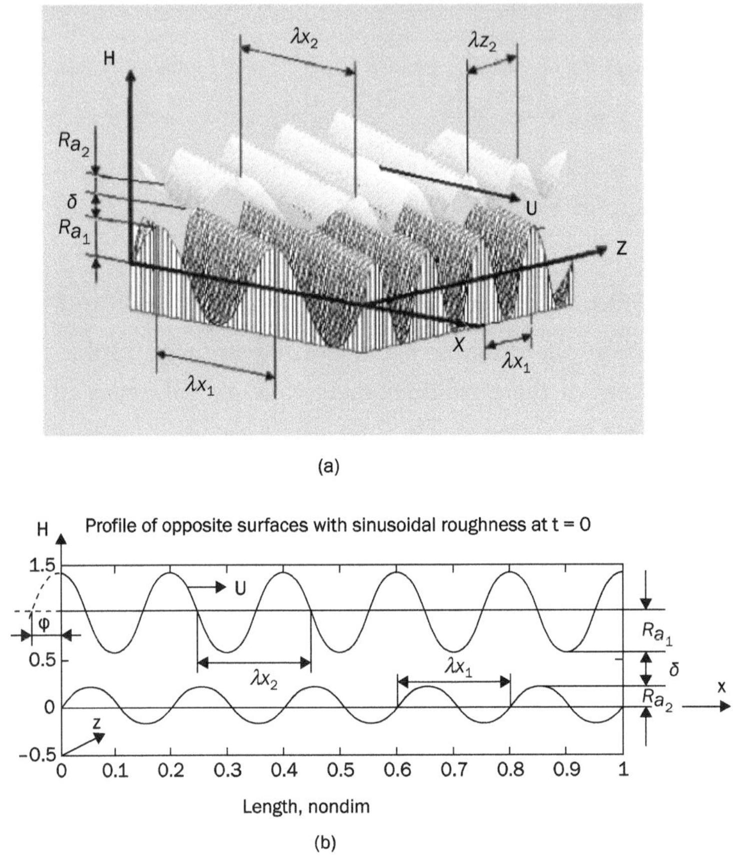

Solution for equal wave numbers

In the case of identical wavelength characteristics in both the X- and Z-directions for the two surfaces, $\ell_1=\ell_2=\ell$, and bearing in mind the trigonometric expressions for a general sinusoid (see, for example, Bronshtein et al., 2007) with amplitudes A_1 and A_2, period $2\pi k$, and phases $\Phi_1=0$

and $(\Phi_2 - T)$ for the moving surface, the gap as per eq. [3.27] can be rewritten as

$$h = 1 + A_t \cos\theta, \qquad [3.28]$$

where

$$A_t = \begin{cases} \sqrt{A_1^2 + A_2^2 - 2A_1A_2\cos 2\pi k(\Phi_2 - T)} & \text{at } A_1 \neq A_2 \\ 2A\sin \pi k\Phi_2 - T & \text{at } A_1 = A_2 \end{cases} \qquad [3.29]$$

$$\theta = 2\pi k(X + Z + \Phi_t) - \pi/2 \qquad [3.30]$$

and

$$\Phi_t = \begin{cases} \dfrac{1}{2\pi k} a\tan\left[\dfrac{A_2 \sin 2\pi k(\Phi_2 - T)}{A_2\cos 2\pi k(\Phi_2 - T) - A_1}\right] & \text{at } A_1 \neq A_2 \\ -\dfrac{1}{4k} + \dfrac{1}{2}(\Phi_2 - T) & \text{at } A_1 = A_2 \end{cases} \qquad [3.31]$$

Here A_t and Φ_t represent the correlated roughness and correlated phase angle respectively. For the case of equal roughnesses $A_1 = A_2 = A$, the trigonometric relations $2\sin^2\dfrac{\alpha}{2} = 1 - \cos\alpha$, $\cos\alpha = \sqrt{1 - \sin^2\alpha}$ were used.

The derivatives on the right-hand side of eq. [3.26] can be rewritten in terms of θ (eq. [3.30]) as:

$$\frac{\partial H}{\partial X} + 2\frac{\partial H}{\partial T} = \frac{\partial H}{\partial X} + 2A_t'\cos\theta + 2\Phi_t'\frac{\partial H}{\partial X}$$

$$= 2A_t'\cos\theta + \left(1 + 2\Phi_t'\right)\frac{\partial H}{\partial X}$$

$$= 2A_t'\cos\theta + 2\pi k\left(1 + \Phi_t'\right)\frac{\partial H}{\partial \theta}$$

where $\Phi_t' = \dfrac{\partial \Phi_t}{\partial T}$, $A_t' = \dfrac{\partial A_t}{\partial T}$.

Bearing in mind the chain rule $\frac{\partial}{\partial x}=\frac{\partial}{\partial \theta}\frac{\partial \theta}{\partial x}, \frac{\partial}{\partial y}=\frac{\partial}{\partial \theta}\frac{\partial \theta}{\partial y}$, eq. [3.26] can evidently be reduced to quasi-one-dimensional form, namely to the ordinary differential equation

$$\frac{\partial}{\partial \theta}\left(H^3\frac{\partial P}{\partial \theta}\right)=\frac{A_t'\cos\theta}{8(\pi k)^2}+\frac{1+2\Phi_t'}{4\pi k}\frac{\partial H}{\partial \theta} \qquad [3.32]$$

This operation is known in mathematical physics as the 'travelling-wave' solution (see, e.g., Polyanin et al., 2006), and represents transition from the two- and three-dimensional relations of types $\omega(x,y)$, $\omega(x,t)$, $\omega(x,y,t)$ to the one-dimensional relation $W(\theta)$, in the case of the linear $\theta(x,y,t)$ form.

Expression [3.32] contains the derivatives of the correlated roughness A_t and phase angle Φ_t; differentiating eqs. [3.29] and [3.31], we have

$$A_t'=\begin{cases}-2\pi k\dfrac{A_1A_2}{A_t}\sin 2\pi k(\Phi_2-T) & \text{at } A_1\neq A_2\\ -2A\cos\pi k(\Phi_2-T) & \text{at } A_1=A_2\end{cases}$$
$$\Phi_t'=\begin{cases}\dfrac{A_2}{A_t^2}\left[A_1\cos 2\pi k(\Phi_2-T)-A_2\right] & \text{at } A_1\neq A_2\\ -\dfrac{1}{2} & \text{at } A_1=A_2\end{cases} \qquad [3.33]$$

Now, double integration of eq. [3.32] with respect to θ yields

$$P=-\frac{A_1A_2\sin 2\pi k(\Phi_2-T)}{2\pi kA_t}\int\frac{\sin\theta d\theta}{H^3}+\frac{A_t^2+2A_2\left[A_1\cos 2\pi k(\Phi_2-T)-A_2\right]}{4\pi kA_t^2}\int\frac{d\theta}{H^2}+C_1\int\frac{d\theta}{H^3}+C_2 \qquad [3.34]$$

where C_1 is an integration constant invariant in θ, and C_2 represents the pressure at the initial θ, also invariant in θ.

The integrals in this solution are the same journal-bearing integrals (Booker, 1965) as in subsection 3.5.2 (cylindrical journal bearing).

For determining the integration constants we need a specific body geometry and possible real pressure values assigned at the boundaries. Consider here boundary conditions as follows

$$\left.\begin{array}{l} P(X=0,Z,T)=P(X=1,Z,T) \\ p(X,Z=0,T)=p(X,Z=1,T)=0 \end{array}\right\}$$

meaning that a divergent region of pressures is assumed, that the opposite boundaries have the same pressures, and that the hydrodynamic effect vanishes on the vertical boundaries. For the lower and upper θ values, we have

$$\begin{array}{l} \theta_1=\theta_{X=0}=\pi k(2Z-T+\Phi_2) \\ \theta_u=\theta_{X=1}=\pi k\left[2(1+Z)-T+\Phi_2\right] \end{array} \qquad [3.35]$$

which represent a periodic pattern with cycles $\theta_1-\theta_u=2\pi k$; l and u denote the lower and upper integration limits respectively.

Subject to the assigned boundary conditions, C_1 can be determined applying the eqs. [3.35] to eq. [3.34], taking the definite integrals between $\theta_l(X=0)$ and $\theta_u(X=1)$, and assuming $p(\theta_u)-C_2=0$ as in the case of a cylindrical bearing. Further, substituting this expression and then determining C_2 from the boundary condition $P=0$ at $Z=0$ associated with

$$\theta=\theta_c=2\pi k\left(X+\Phi_t-\frac{1}{4k}\right), \qquad [3.36]$$

we have an expression for $P(\theta)$ which, after the requisite transformations, reads

$$P=\left\{A_t^2+2A_2\left[A_1\cos 2\pi k(\Phi_2-T)-A_2\right]\right\}$$
$$\left[\sin\theta(2+A_t\cos\theta)(1+A_t\cos\theta_C)^2\right. \qquad [3.37]$$
$$\left.-\sin\theta_C(2+A_t\cos\theta_C)(1+A_t\cos\theta)^2\right]\Big/$$
$$4\pi kA_t(2+A_t^2)(1+A_t\cos\theta_C)^2(1+A_t\cos\theta)^2$$
$$+\frac{A_1A_2\sin 2\pi k(\varphi_2-t)(\cos\theta_C-\cos\theta)\left[2+A_t(\cos\theta_C+\cos\theta)\right]}{4\pi kA_t(1+A_t\cos\theta_C)^2(1+A_t\cos\theta)^2}$$

In the particular case of equal asperity heights $A_1=A_2=A$ we have $A_t^2=2A^2\left[1-\cos 2\pi k(\Phi_2-T)\right]$, the first coefficient in the first term of eq. [3.37] vanishes and the pressure equation reduces to

$$P=-A\sin 2\pi k(\Phi_2-T)(\cos\theta_C-\cos\theta)$$
$$\left[1+A\sin\pi k(\Phi_2-T)(\cos\theta_C+\cos\theta)\right]\Big/$$
$$\pi k(1+2A\sin\pi k(\Phi_2-T)\cos\theta_C)^2 \qquad [3.38]$$
$$(1+2A\sin\pi k(\Phi_2-T)\cos\theta)^2$$

Eqs. [3.37] and [3.38] can be reconverted from the variable θ to the original ones – *X*, *Z* and *T* – by substitution of expressions [3.30] and [3.36] for θ and θ_C; the final expression is very long and is not given here (available in Burstein, 2010); the *P*, θ, and θ_C expressions constitute a complete algebraic set of equations which suffices for routine arithmetic calculations. Typical *X*, *Z* – and *X*, *T* – pressure distributions for different roughnesses were calculated with the aid of these equations and are given in Fig. 3.11(a) and (b). As the pressure distribution is periodical, its examination for one wavelength yields complete information about the surface behaviour as a whole.

Figure 3.11 Typical pressure distribution in lubricating film between sinusoidal surfaces with wave number 2, wave ratio 1, and asperity height of lower surface 0.5 and upper surface 0.25, at time 1/(2*k*) (a) and at coordinate y = 1/(2*k*) (b).

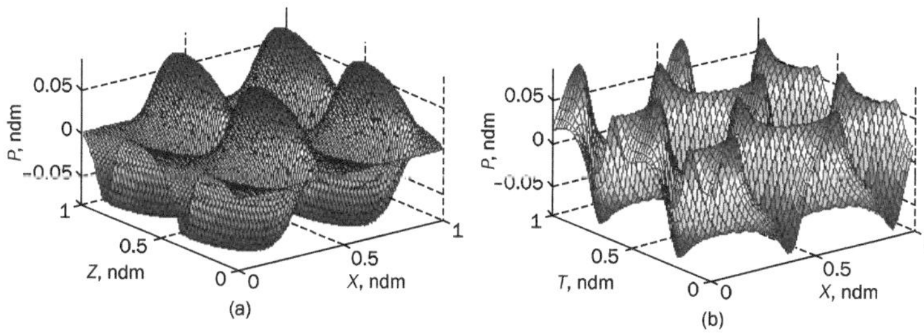

Solution for unequal wave numbers

The above relations can be extended to the case of unequal wave numbers along axes X and Z, wave number ratios $\ell_1 = \ell_2 = \ell \neq 1$. In this case the gap equation has the form [3.28] with A_t as per eq. [3.29] and θ given by a relation in which ℓ does not equal 1:

$$\theta = 2\pi k\left(X + \ell^{-1}Z + \Phi_t - \frac{1}{4k}\right) \qquad [3.39]$$

Reynolds' equation for this case reads

$$\left(1+\ell^{-2}\right)\frac{\partial}{\partial\theta}\left(H^3\frac{\partial P}{\partial\theta}\right) = \frac{A_t'\cos\theta}{2(\pi k)^2} + \frac{1+2\Phi_t'}{2\pi k}\frac{\partial H}{\partial\theta} \qquad [3.40]$$

Integrating this equation successively as in the preceding case, and determining the integration constants from the same boundary conditions, we easily obtain an analogue of eq. [3.37] that is valid for the case of different asperity heights and different wave numbers in the X- and Z-directions

$$P=\left\{A_t^2+2A_2\left[A_1\cos 2\pi k\left(\Phi_2-T\right)-A_2\right]\right\}$$
$$\times\left[\sin\theta\left(2+A_t\cos\theta\right)\left(1+A_t\cos\theta_c\right)^2\right.$$
$$\left.-\sin\theta_c\left(2+A_t\cos\theta_c\right)\left(1+A_t\cos\theta\right)^2\right]\Big/ \qquad [3.41]$$
$$2\pi kA_t\left(2+A_t^2\right)\left(1+\ell^{-2}\right)\left(1+A_t\cos\theta_c\right)\left(1+A_t\cos\theta\right)^2$$
$$+\frac{A_1A_2\sin 2\pi k\left(\Phi_2-T\right)\left(\cos\theta_c-\cos\theta\right)^2\left[2+A_t\left(\cos\theta_c+\cos\theta\right)\right]}{2\pi kA_t\left(1+\ell^{-2}\right)\left(1+A_t\cos\theta_c\right)^2\left(1+A_t\cos\theta\right)^2}$$

with θ as per eq. [3.39] and θ_c defined by

$$\theta_C=2\pi k\left(\ell^{-1}Z+\Phi_t-\frac{1}{4k}\right) \qquad [3.42]$$

Eq. [3.41] is more general than eq. [3.37], being valid for both $\ell=1$ and $\ell\neq 1$.

Typical pressure-coordinate and pressure-coordinate-time distributions are presented in Fig. 3.12.

It is seen from Fig. 3.12(a), wave ratio 0.5, that the wave number in the X-direction is 2 and in the Z-direction 4. The waves along $Z=0$ are partly truncated due to displacement

Figure 3.12 **Typical pressure distribution in lubricating film between rough surfaces with wave number 2 at wave ratio 0.5 with asperity height of lower surface 0.5 and of upper surface 0.25 along: (a) entire rough surface at time 1/(2*k*); (b) *X*, *T* coordinates at *Z*=1/(4*k*)**

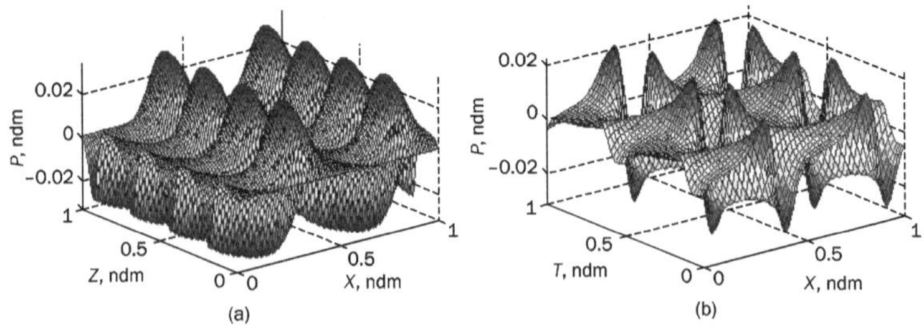

of the moving-surface origin by a phase angle $1/(4k)$. The time dependence, Fig. 3.12(b), has a more complicated form, but it should be borne in mind that the plot refers to a single Z-point and animated pictures are needed for studying the overall situation. A feature common to the coordinate and time pressure distributions is asymmetry about the X,Z – and X,T – planes. This fact and the relevant theoretical relations were verified by numerical calculations and are described in the next section.

Effect of asperity height and intra-surface wave ratio on hydrodynamic pressure

The effect of sinusoidal roughness on the pressure distribution and normal load support was studied with data presented in Table 3.1. The rather large reference values of the viscosity were adopted for a better hydrodynamic effect on the resulting pressure data.

Table 3.1 **Reference values applied in calculations**

Parameter	Notation	Reference value	Dimensionality
Sliding velocity	u	8.9	m/s
Fluid viscosity	μ	3, 0.3	Pa·s
Clearance	Δ	6	μm
Surface sizes	L_x, L_z	10	mm
Asperity height of lower surface	R_{a_1}	1.5 . . . 6	μm
Asperity height of upper surface	R_{a_2}	3	μm
Relative gap	$H_0=\Delta+2R_{a_2}$	12	μm
Inner-surface wave ratio	ℓ	1, 2	dimensionless
Number of profile waves	k	100 . . . 35,000	units

As it is obviously impossible to study all instantaneous pressures, the maximum value is usually taken for this purpose. It has a strong influence on normal film support and on such an undesirable effect as cavitation. The latter occurs when the hydrodynamic pressure reaches, or drops below, its normal level and is characterized by decomposition of lube into fractions, bubbling and formation of cavities on the surface. The adopted reference values of the dynamic viscosity were those of motor oils SAE 30–50 at 100 °C. The dimensionles cavitation pressure threshold P_{cav} equals $0.91 \cdot 10^{-5}$ at viscosity 3 Pa·s, and $9.1 \cdot 10^{-3}$ at 0.3 Pa·s.

Figure 3.13 shows the dependence of the maximal positive hydrodynamic pressure for the asperity height ratio range 0.5–2 and for wave ratios 0.5, 1 and 2. Evidently, the pressure is lower for lower wave ratios; besides, as could be expected,

Figure 3.13 **Maximal pressures versus asperity height ratio at different wave ratios with reference wave number $k_x = 100$**

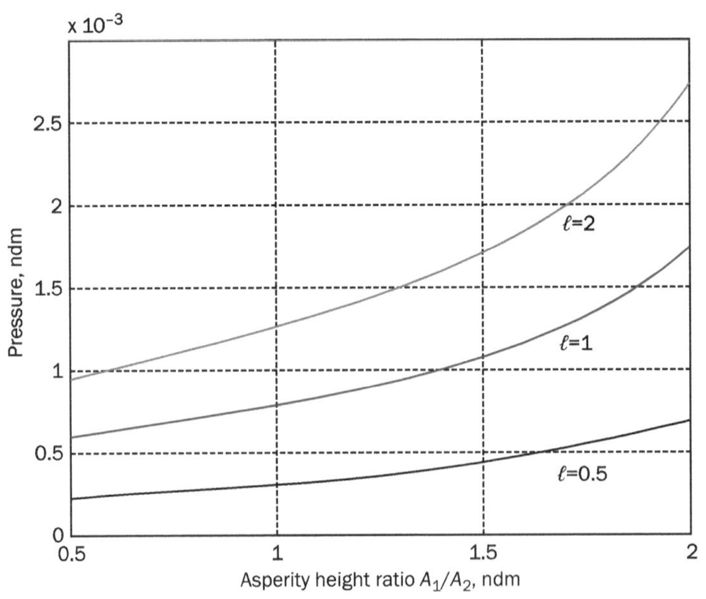

if the wave number in the X-direction (that of motion) is higher than in the Z-direction (wave ratio 0.5), the maximal pressure is double and more than in the opposite case (wave ratio 2). As can be seen, the maximal support effect is reached at larger values of the roughness height ratio A_1/A_2, that is, the roughness of the moving surface affects the hydrodynamic pressure more strongly than that of the stationary one. Also, maximal pressures are reached at the largest values of wave ratio ℓ, which means the wave number in the X-direction outweighs that in the Z-direction.

Figure 3.14 shows the dependence of the maximal pressure on the wave number for asperity height ratio range 0.5–2 at $\ell=1$ and $\ell=2$. Numerical results are also presented in Table 3.2. As can be expected, higher pressures are reached at smaller wave numbers. Thus it is preferable to produce surfaces

Figure 3.14 Maximal and cavitation pressures at different numbers of waves and roughness values at wave ratio 1 (a) and 2 (b)

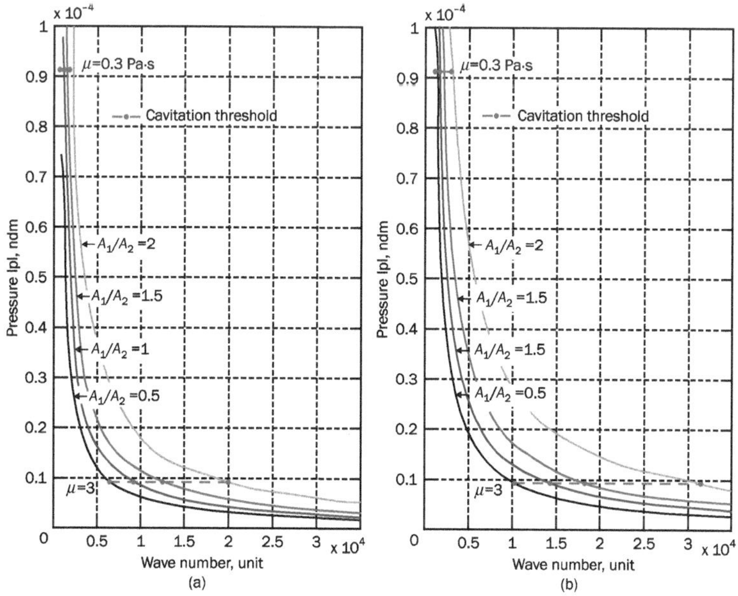

Table 3.2 **Cavitation wave number at studied values of viscosity, wave and asperity height ratios**

Wave ratio	Viscosity, Pa·s	$A_1/A_2=0.5$	$A_1/A_2=1$	$A_1/A_2=1.5$	$A_1/A_2=2$
1	0.3	628	919	1302	1824
	3.0	6545	8571	11754	19534
2	0.3	1143	1496	1847	3012
	3.0	10297	13824	19020	30964

with fewer waves, but unfortunately unrealistic, as in practice the surfaces of machine parts have thousands of roughness waves.

The horizontal dotted lines in Fig. 3.14(a) and (b) represent the two cavitation thresholds for two viscosity values – 3 (lower threshold line) and 0.3 Pa·s. Below these lines the positive hydrodynamic effect did not exist at the respective viscosities.

As can be seen, for wave ratio 1 (Fig. 3.14(a)) and viscosity 3 Pa·s the hydrodynamic effect vanished at wave numbers above $6.5 \cdot 10^3$ at the lower roughness height ratio 0.5 and about $19.5 \cdot 10^3$ at the higher ratio 2. For the lower viscosity 0.3 Pa·s the effect vanished at wave numbers about ten times lower – from above $0.63 \cdot 10^3$ and $1.8 \cdot 10^3$ for the two ratios respectively. For the unequal wave numbers, in the higher case (Fig. 14(b) – wave ratio 2), the cavitation threshold was reached at higher wave numbers, which means that the surface apparently can contain more waves and survive under this maximal pressure. The hydrodynamic effect vanished at higher wave numbers; specifically, for viscosity 3 Pa·s it was above $10.3 \cdot 10^3$ at asperity height ratio 0.5 and above $31.0 \cdot 10^3$ at asperity height ratio 2; for viscosity 0.3 Pa·s, cavitation set in when the surface had about $1.1 \cdot 10^3$ waves at ratio 0.5 and $3.0 \cdot 10^3$ waves at asperity ratio 2.

Thus we can say in general that a positive effect takes place only at wave numbers below 600 at all studied roughness ratios. As in most real oils the viscosities are ten and more times lower than the discussed values, and the cavitation threshold can set in with two-digit or even one-digit numbers waves. This explains why real mechanical parts do not exhibit any positive hydrodynamic effect. Most such parts are sufficiently large to accommodate tens and hundreds of thousands of wavelengths, so that a sinusoidal profile is irrelevant to them, but for small ones, such as micro- or nano-scale lubricated surfaces, it can be very effective.

The discussed results were computed by a special program written in Matlab. The algorithm and other information about this program are available in Burstein (2009). Summarizing this analysis, the main results are:

- The maximal support effect is reached at larger values of the roughness height ratio A_1/A_2, indicating that the roughness of the stationary surface has a stronger effect on the hydrodynamic pressure than that of the moving one; this is of practical interest in designing opposite surfaces with different roughnesses.
- In the case of equal wave numbers in the X- and Z-directions, for the higher viscosity 3 Pa·s the hydrodynamic effect vanishes at a wave number above $6.5 \cdot 10^3$ at roughness height ratio 0.5 and at about $19.5 \cdot 10^3$ at the higher ratio 2. For the lower viscosity 0.3 Pa·s the hydrodynamic effect vanishes at wave numbers about ten times lower – from above $0.63 \cdot 10^3$ to $1.8 \cdot 10^3$ for asperity ratios 0.5 and 2 respectively.
- In the case of unequal wave numbers, the hydrodynamic effect vanishes for 3Pa·s at above $10.3 \cdot 10^3$, roughness ratio 0.5, and at above $31.0 \cdot 10^3$, roughness height ratio 2. For 0.3 Pa·s it vanishes when the surface has about $1.1 \cdot 10^3$ waves at roughness height ratio 0.5 and $3.0 \cdot 10^3$ waves at roughness height ratio 2.

- In general, the maximal pressures are reached at larger wave ratios, indicating that the wave number in the X-direction outweighs that in the Z-direction.

The results in this section were obtained for the case of equal wave numbers on the surfaces. The 'unequal' case was, until quite recently, considered intractable because of the serious difficulties in integrating Reynolds' equation (see, e.g., Letallear et al., 2002). This obstacle was removed on introduction of the triangular roughness pattern.

3.6.3 Hydrodynamic lubrication for triangular roughness

Take for analysis a control cell – an element on the surfaces such that integer numbers of waves are located along the coordinate axes. By this means the behaviour of the whole surface can be studied on the basis of a small part. Thus a surface of, say, 10 sq.cm. with asperity width 50 micron (roughness standard numbers N8–N12) has approximately 630 asperities in each of the X- and Z-directions, but one can make do with a cell with say 5 asperities only, and apply the results to all 126 such cells that can be accommodated on the surface; in parallel, the opposite surface may have a different asperity width, but again such that the wave numbers in both directions are integers; in this example, for say 3 waves, the asperity width must be about 83micron. It is also clear that in reality the number of asperities in the control cell should be varied over a relatively narrow range, with the wave numbers on the two surfaces varying by a factor of 2 or 3.

The X-plane of the profile geometry and the 3D view for a control cell with unequal wave numbers and unequal asperity heights, R_a, are shown in Fig. 3.15 for triangular waves, as an example.

Figure 3.15 Schematic of surface (a), surface profile (b), and gap geometry (c) at wave numbers $k_{1_x}=3$, $k_{2_x}=5$, $k_{1_z}=2$, $k_{2_z}=3$, and time and phase displacement of upper surface $\Phi=T=1/(2\ k_{2x})$

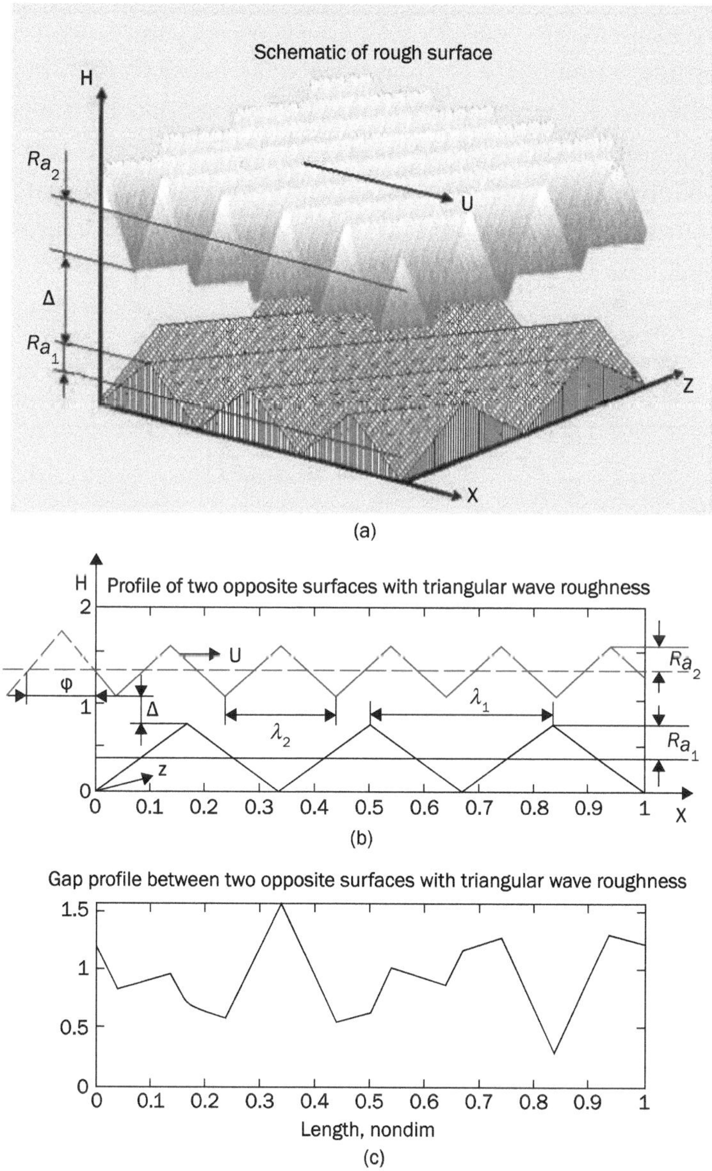

For a pair of surfaces as in Fig. 3.15 with surface profiles $h^{(1)}(x,z)$ and $h^{(2)}(x,z,t)$, the film thickness between them is

$$h(x,z,t) = h^{(2)}(x,z,t) - h^{(1)}(x,z) \qquad [3.43]$$

In the studied general case we have triangular waves with: $k_{1_x} \neq k_{2_x} \neq k_{1_z} \neq k_{2_z}$ and $R_{a_1} \neq R_{a_2}$. Such waves can be described by a function of the form asin(sin α) or acos(cos β)/π. In the latter case a pythagorean number is introduced for the amplitude, varying in the interval 0–1.

Considering all the above and taking the origin of coordinates at the point where the phase angle $\varphi_1 = 0$, the film thickness [3.43] can be written as

$$h = \Delta + 2R_{a_1} + \frac{2R_{a_2}}{\pi} \mathrm{a}\cos\cos\left[\frac{2\pi k_{2_x}}{L_x}\left(x + \ell_S \ell_{k_2} z - ut\right)\right]$$
$$- \frac{2R_{a_1}}{\pi} \mathrm{a}\cos\cos\left[\frac{2\pi k_{1_x}}{L_z}\left(x + \ell_S \ell_{k_1} z\right)\right] \qquad [3.44]$$

Here for simplicity the initial roughness phase shift φ_2 of the upper surface relative to the lower is given implicitly by the time t; $\ell_{k_1} = k_{z_1}/k_{x_1}$ and $\ell_{k_2} = k_{z_2}/k_{x_2}$ represent the wave ratios for the lower and upper surfaces. In dimensionless form eq. [3.44] reads

$$H = 1 + A_2 \mathrm{a}\cos\cos 2\pi k_{2_x}\left(X + \ell_S \ell_{k_2} Z - T\right)$$
$$- A_1 \mathrm{a}\cos\cos 2\pi k_{1_x}\left(X + \ell_S \ell_{k_1} Z\right) \qquad [3.45]$$

where $A_1 = R_{a_1}/[\pi\ (\Delta/2 + R_{a_1})]$, $A_2 = R_{a_2}/[\pi\ (\Delta/2 + R_{a_1})]$, and X and Y as at [3.26].

For the current case the dimensionless Reynolds' equation as per [3.26] with the same boundary conditions as in the sinusoidal roughness case, should be used.

Next assume $\ell_{k_1} = \ell_{k_2} = \ell_k$, which requires equality of the wave ratios for the two surfaces, $k_{1_Z}/k_{1_X} = k_{2_Z}/k_{2_X}$, but admits different wave numbers for them. Thus for $k_{1_x}=2$, $k_{2_x}=3$, $k_{1_z}=4$, and $k_{2_z}=6$ we have $k_{1_Z}/k_{1_X}=k_{2_Z}/k_{2_X}=2$. Further, the travelling wave transform is introduced similar to that in subsection 3.6.1:

$$\begin{aligned}
&\theta = X + \ell_S \ell_k Z \\
&\frac{\partial H}{\partial X} = \frac{\partial H}{\partial \theta} \\
&\frac{\partial P}{\partial X} = \frac{\partial P}{\partial \theta} \\
&\frac{\partial P}{\partial Z} = \frac{\partial P}{\partial \theta}\frac{\partial \theta}{\partial Z} = \ell_S \ell_k \frac{\partial P}{\partial \theta}
\end{aligned} \qquad [3.46]$$

Thus for the film thickness

$$H = 1 + A_2 a \cos\cos 2\pi k_{2_x}(\theta - T) - A_1 a \cos\cos 2\pi k_{1_x}\theta \qquad [3.47]$$

and the 2D Reynolds equation is reduced to 1D

$$\left(1+\ell_S^2\ell_k^2\right)\frac{\partial}{\partial\theta}\left(H^3\frac{\partial P}{\partial\theta}\right) = \frac{\partial H}{\partial\theta} + 2\frac{\partial H}{\partial T} \qquad [3.48]$$

The boundary conditions reduce to $P=0$ at $\theta=0$ and $\theta=1+\ell_S\ell_k$, and P is identical at $\theta=X$ and $\theta=X+\ell_S\ell_k$. The solution of Reynolds' equation should comply with these conditions.

Theoretical solution

For further manipulation of eq. [3.31] we need to determine all X-values at which the waves' peaks or hollows occur. For $H_1(\theta)$, eq. [3.47], these points correspond to $\sin 2\pi k_{1_x}\theta = 0$, and for $H_2(\theta)$ to $\sin 2\pi k_{2_x}(\theta - T) = 0$. Thus for the lower wave $2\pi k_1 \theta_e = \pi(n_1 - 1)$, and the extreme points are

$$\begin{aligned}
&\theta_{e_1} = \frac{n_1 - 1}{2k_{1_x}} \\
&X_{e_1} = \theta_{e_1} - \ell_S \ell_k Z
\end{aligned} \qquad [3.49]$$

The second expression in [3.49] should be used in the case of the X and Z coordinates. Analogously, for the upper wave $2\pi k_2\left(\theta_{e_2}-T\right)=\pi\left(n_2-1\right)$, and the extreme points are

$$\begin{aligned}\theta_{e_2}&=\frac{n_2-1}{2k_{2_x}}+T\\X_{e_2}&=\theta_{e_2}-\ _{S\ k}Z+T\end{aligned}\qquad[3.50]$$

where $n_1=1,2,\ldots,2k_{1_x}$ and $n_2=1,2,\ldots,2k_{2_x}$. Thus the gap contour has $m = n_1+n_2$ break points with coordinates X_{e_1} and X_{e_2}, and $m+1$ segments. Note that eqs. [3.49] and [3.50] cover both the inner and boundary points.

For solving eq. [3.48] one needs to find the differentials on its right-hand side and then differentiate it twice. Differentiating H, we have after some transformations:

with respect to θ

$$\frac{\partial H}{\partial\theta}=2\pi A_2\delta_{A_2}k_2\left(1-\delta_A\ell_A\ell\right)\qquad[3.51]$$

and with respect to T

$$\frac{\partial H}{\partial T}=-2\pi A_2\delta_{A_2}k_2\qquad[3.52]$$

where

$$\begin{aligned}\delta_{A_1}&=\begin{cases}-1 \text{ at } \sin 2\pi k_1\theta<0\\ \ \ 1 \text{ at } \sin 2\pi k_1\theta\geq 0\end{cases}\\\delta_{A_2}&=\begin{cases}\ \ 1 \text{ at } \sin 2\pi k_2\left(\theta-T\right)\leq 0\\-1 \text{ at } \sin 2\pi k_2\left(\theta-T\right)>0\end{cases}\\\delta_A=\delta_{A_1}/\delta_{A_2}&=\begin{cases}-1 \text{ at } \sin 2\pi k_1\theta\sin 2\pi k_2\left(\theta-T\right)\leq 0\\ \ \ 1 \text{ at } \sin 2\pi k_1\theta\sin 2\pi k_2>0\end{cases}\end{aligned}\qquad[3.53]$$

$\ell_A=A_1/A_2$ – asperity height ratio, and $\ell=k_1/k_2$ – inter-surface wave ratio.

By eqs. [3.51] and [3.52] the right member of eq. [3.48] reads

$$\frac{\partial H}{\partial \theta}+2\frac{\partial H}{\partial T}=-2\pi A_2 \delta_{A_2} k_2 \left(1+\delta_A \ell_A \ell\right) \qquad [3.54]$$

whence

$$\left(1+\ell_S^2 \ell_k^2\right)\frac{\partial}{\partial \theta}\left(H^3 \frac{\partial P}{\partial \theta}\right)=-2\pi A_2 \delta_{A_2} k_2 \left(1+\delta_A \ell_A \ell\right) \qquad [3.55]$$

The right member in eq. [3.55] is piecewise constant along the segments of the gap described by [3.47]. For simplicity, we replace the variable θ in [3.55] by H via the chain rule

$$\frac{\partial}{\partial \theta}=\frac{\partial}{\partial H}\frac{\partial H}{\partial \theta}=2\pi A_2 \delta_{A_2} k_2 \left(1-\delta_A \ell_A \ell\right)\frac{\partial}{\partial H} \qquad [3.56]$$

and obtain for each i-th segment the following expression

$$\left(1+\ell_S^2 \ell_k^2\right)\frac{\partial}{\partial H}\left(H^3 \frac{\partial P}{\partial \theta}\right)=-\frac{1+\delta_A \ell_A \ell}{1-\delta_A \ell_A \ell} \qquad [3.57]$$

Accordingly, eq. [3.57] should be integrated over each segment (a very simple operation) and the resulting integral normalized against that of its predecessor. After the first integration

$$H^3 \frac{\partial P}{\partial \theta}=\frac{1}{1+\ell_S^2 \ell_k^2}\left[-\frac{1+\delta_A \ell_A \ell}{1-\delta_A \ell_A \ell}H+C_1^{(i)}\right] \qquad [3.58]$$

where $C_1^{(i)}$ is the integration constant for the segment in question.

Re-applying the chain rule

$$H^3 \frac{\partial P}{\partial H} = -\frac{1}{2\pi A_2 \delta_{A_2} k_2 (1-\delta_A \ell_A \ell)\left(1+\ell_S^2 \ell_k^2\right)} \left[\frac{1+\delta \ell_A \ell}{1-\delta \ell_A \ell} H - C_1^{(i)}\right] \qquad [3.59]$$

After the second integration we have

$$P = C_2^{(i)} - \frac{1}{2\pi A_2 \delta_{z_2} k_2 (1-\delta_A \ell_A \ell)\left(1+\ell_s^2 \ell_k^2\right)} \left[\frac{1+\delta_A \ell_A \ell}{1-\delta_A \ell_A \ell} \int \frac{1}{H^2} dH - C_1^{(i)} \int \frac{1}{H^3} dH\right] \qquad [3.60]$$

and finally arrive at the equation for the pressure distribution:

$$P = C_2^{(i)} + \frac{1}{2\pi A_2 \delta_{A_2} k_2 (1-\delta_A \ell_A \ell)\left(1+\ell_s^2 \ell_k^2\right)} \left[\frac{1+\delta_A \ell_A \ell}{1-\delta_A \ell_A \ell} \frac{1}{H} - C_1^{(i)} \frac{1}{2H^2}\right] \qquad [3.61]$$

In the case of $k_2 = k_1$ and $A_1 = A_2$, eq. [3.61] has singularities at points where $\delta_{A_1} = \delta_{A_2} = +1 \cap -1$ and is unsuitable, but still applicable in all cases (at the piecewise segments of the gap) in which even one of these conditions is not observed.

The constants of integration have to be determined for each of the m + 1 segments. This is done via two conditions – the pressures and pressure gradients should coincide at m critical (extreme) points reached from the left (i-th segment) and from the right [(i + 1)-th segment]:

$$\left.\begin{aligned} P_e^{(i+1)} &= P_e^{(i)}, \\ \left.\frac{\partial P}{\partial \theta}\right|_e^{(i+1)} &= \left.\frac{\partial P}{\partial \theta}\right|_e^{(i)} \end{aligned}\right\} \qquad [3.62]$$

Applying eq. [3.62] to eqs. [3.59] and [3.61], we obtain

$$C_1^{(i+1)} - C_1^{(i)} = \left(\frac{1+\delta_A^{(i+1)}\ell_A\ell}{1-\delta_A^{(i+1)}\ell_A\ell} - \frac{1+\delta_A^{(i)}\ell_A\ell}{1-\delta_1^{(i)}\ell_A\ell} \right) H_i \qquad [3.63]$$

and

$$\begin{aligned} & C_2^{(i+1)} - C_2^{(i)} + \frac{1}{4\pi A_2 k_2 \left(1+\ell_s^2\ell_k^2\right) H_i^2} \\ & \times \left[\frac{C_1^{(i+1)}}{\delta_{A_2}^{(i+1)}\left(1-\delta_A^{(i+1)}\ell_A\ell\right)} - \frac{C_1^{(i)}}{\delta_{A_2}^{(i)}\left(1-\delta_A^{(i)}\ell_A\ell\right)} \right] \\ & = -\frac{1}{2\pi A_2 k_2 \left(1+\ell_s^2\ell_k^2\right) H_i} \\ & \times \left[\frac{1+\delta_A^{(i+1)}\ell_A\ell}{\delta_{A_2}^{(i+1)}\left(1-\delta_A^{(i+1)}\ell_A\ell\right)^2} - \frac{1+\delta_A^{(i)}\ell_A\ell}{\delta_{z_2}^{(i)}\left(1-\delta_A^{(i)}\ell_A\ell\right)^2} \right] \end{aligned} \qquad [3.64]$$

Thus there are 2m equations and 2(m + 1) constants C_1 and C_2. The missing two equations are obtainable from the boundary conditions:

$$\begin{aligned} & -C_1^{(1)} \frac{1}{4\pi A_2 \delta_{A_2}^{(1)} k_2 \left(1-\delta_A\ell_A\ell\right)\left(1+\ell_S^2\ell_k^2\right) H_1^2} + C_2^{(1)} \\ & = -\frac{1+\delta_A^{(1)}\ell_A\ell}{\left(1-\delta_A^{(1)}\ell_A\ell\right)^2} \frac{1}{2\pi A_2 \delta_{A_2}^{(1)} k_2 \left(1+\ell_S^2\ell_k^2\right) H_1} \end{aligned}$$

and

$$-C_1^{(m+1)}\frac{1}{4\pi A_2\delta_{A_2}^{(m+1)}k_2\left(1+\ell_s^2\ell_k^2\right)\left(1-\delta_A^{(m+1)}\ell_A\ell\right)H_{m+2}^2}+C_2^{(m+1)}$$
$$=-\frac{1+\delta_A^{(m+1)}\ell_A\ell}{\left(1-\delta_A^{(m+1)}\ell_A\ell\right)^2}\frac{1}{2\pi A_2\delta_{A_2}^{(m+1)}k_2\left(1+\ell_s^2\ell_k^2\right)H_{m+2}}$$

[3.65]

In matrix form, the equations read **DC** = **B** where the **D** is a square matrix

$$\mathbf{D}=\begin{bmatrix}
-1 & 1 & 0 & \cdots & 0 & 0 & 0 & 0 & 0 & \cdots & 0 & 0\\
0 & -1 & 1 & \cdots & 0 & 0 & 0 & 0 & 0 & \cdots & 0 & 0\\
\cdots & \cdots & \cdots & \cdots & \cdots & \cdots & \cdots & \cdots & \cdots & \cdots & \cdots & \cdots\\
0 & 0 & 0 & \cdots & -1 & 1 & 0 & 0 & 0 & \cdots & 0 & 0\\
\frac{K_1}{2H_2^2} & -\frac{K_2}{2H_2^2} & 0 & \cdots & 0 & 0 & -1 & 1 & 0 & \cdots & 0 & 0\\
0 & \frac{K_2}{2H_3^2} & -\frac{K_3}{2H_3^2} & \cdots & 0 & 0 & 0 & -1 & 1 & \cdots & 0 & 0\\
\cdots & \cdots & \cdots & \cdots & \cdots & \cdots & \cdots & \cdots & \cdots & \cdots & \cdots & \cdots\\
0 & 0 & 0 & \cdots & \frac{K_{m-2}}{2H_{m-1}^2} & -\frac{K_{m-1}}{2H_{m-1}^2} & 0 & 0 & 0 & \cdots & -1 & 1\\
-\frac{K_1}{2H_1^2} & 0 & 0 & \cdots & 0 & 0 & 1 & 0 & 0 & \cdots & 0 & 0\\
0 & 0 & 0 & \cdots & 0 & -\frac{K_{m-1}}{2H_{m+2}^2} & 0 & 0 & 0 & \cdots & 0 & 1
\end{bmatrix}$$

[3.66]

where

$$K_i=\frac{1}{2\pi A_2\delta_{A_2}^{(i)}k_2\left(1-\delta_A^{(i)}\ell_A\ell\right)\left(1+\ell_S^2\ell_k^2\right)} \quad [3.67]$$

and **B** is a column vector as follows

$$
B=\left[\begin{array}{c}
\left.\begin{array}{c}
\left(\dfrac{1+\delta_z^{(2)}\ell_z\ell}{1-\delta_z^{(2)}\ell_z\ell}-\dfrac{1+\delta_z^{(1)}\ell_z\ell}{1-\delta_z^{(1)}\ell_z\ell}\right)H_2\\
\left(\dfrac{1+\delta_A^{(3)}\ell_A\ell}{1-\delta_A^{(3)}\ell_A\ell}-\dfrac{1+\delta_A^{(2)}\ell_A\ell}{1-\delta_A^{(2)}\ell_A\ell}\right)H_3\\
\cdots\\
\left(\dfrac{1+\delta_A^{(m+1)}\ell_A\ell}{1-\delta_A^{(m+1)}\ell_A\ell}-\dfrac{1+\delta_A^{(m)}\ell_A\ell}{1-\delta_A^{(m)}\ell_A\ell}\right)H_{m+1}
\end{array}\right\} m \text{ lines}\\
\left.\begin{array}{c}
-\left(K_2\dfrac{1+\delta_A^{(2)}\ell_A\ell}{1-\delta_A^{(2)}\ell_A\ell}-K_1\dfrac{1+\delta_A^{(1)}\ell_A\ell}{1-\delta_A^{(1)}\ell_A\ell}\right)\dfrac{1}{2H_2}\\
-\left(K_3\dfrac{1+\delta_A^{(3)}\ell_A\ell}{1-\delta_A^{(3)}\ell_A\ell}-K_2\dfrac{1+\delta_A^{(2)}\ell_A\ell}{1-\delta_A^{(2)}\ell_A\ell}\right)\dfrac{1}{2H_3}\\
\cdots\\
-\left(K_{m-2}\dfrac{1+\delta_A^{(m-2)}\ell_A\ell}{1-\delta_A^{(m-2)}\ell_A\ell}-K_{m-1}\dfrac{1+\delta_A^{(m-1)}\ell_A\ell}{1-\delta_A^{(m-1)}\ell_A\ell}\right)\dfrac{1}{2H_{m-1}}\\
-K_1\dfrac{1+\delta_A^{(1)}\ell_A\ell}{1-\delta_A^{(1)}\ell_A\ell}\dfrac{1}{2H_1}\\
-K_{m-1}\dfrac{1+\delta_A^{(m\ 1)}\ell_A\ell}{1-\delta_A^{(m-1)}\ell_A\ell}\dfrac{1}{2H_m}
\end{array}\right\} m \text{ lines}
\end{array}\right]
\qquad [3.68]
$$

The solution is the column vector of the search constants $C_{1,2}^{(i)} = D\backslash B$ in which the first m terms are the search constants $C_1^{(i)}$ and the second m terms are $C_2^{(i)}$.

With all constants determined, eq. [3.61] can be conveniently used for calculating the hydrodynamic pressure at every point of the lubricating film.

Typical pressure-coordinate and pressure-coordinate-time distributions at $A_1 = A_2 = 0.15$ are presented in Fig. 3.16(a) and (b) (inter-surface wave ratio 3/4) and Fig. 3.17(a) and (b) (inter-surface wave ratios 3/4 and 4/3). It is seen from

Figure 3.16 Typical pressure distribution in lubricating film between triangular wave surfaces at asperity height $A_1 = A_2 = 0.15$, and inner-surface wave ratio 3/4, along: (a) entire surface profile at time $1/(k_{2_x})$ and (b) X, T coordinates at $Z = 1/(k_{2_x})$

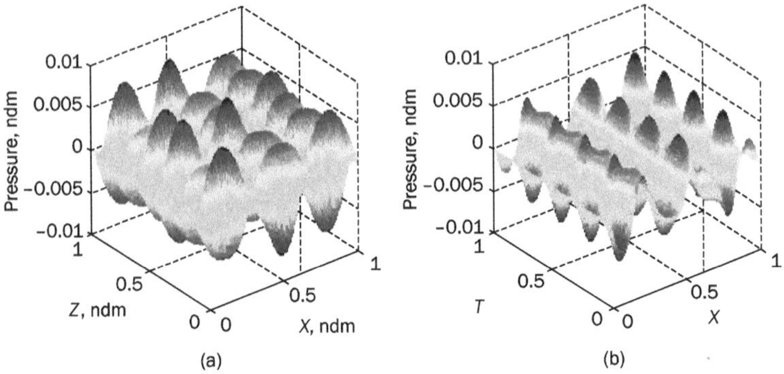

Figure 3.17 Typical pressure distribution in lubricating film between triangular wave surfaces at asperity height $A_1 = A_2 = 0.15$, and inner-surface wave ratio 4/3 along: (a) entire surface profile at time $1/(k_{2_x})$ and (b) X, T coordinates at $Z = 1/(k_{2_x})$

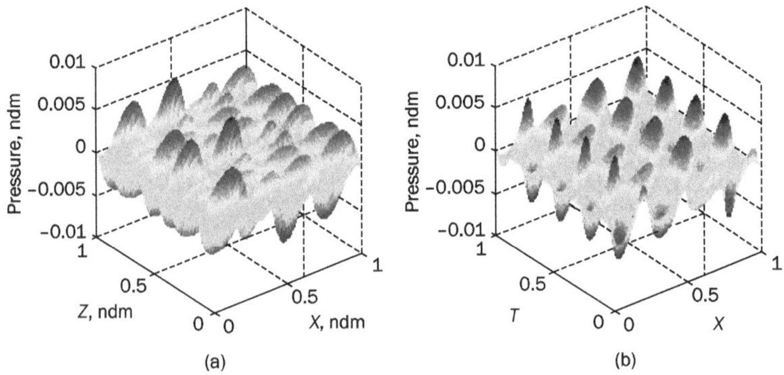

Fig 3.16(a) and Fig. 3.17(a) that the number of maximum/minimum pressure points in the X-direction is 3 and 4, i.e. equal to the wave numbers on the stationary surface, and the same is the case with the Z-direction. The dependence of pressure on time, Fig 3.16(b) and Fig. 3.17(b), exhibits

similar behaviour. But it should be borne in mind that the plots are given for a single time moment T (Figs 3.16(b), 3.17(b)) and a single coordinate point Z (Figs 3.16(a), 3.17(a)); animated pictures are needed for studying the overall situation.

The pressure distribution is not periodically identical along the X-axis (as was the case of sinusoidal roughness with the same wave numbers on both surfaces), while along the T-axis such identity is evidently preserved. Thus examination of a single wavelength does not yield complete information about the surface behaviour as a whole.

Effect of asperity height, inter-surface and intra-surface wave ratios on hydrodynamic pressure

Roughened surfaces with the data of Table 3.3 were studied. The reference data yield to the dimensionless cavitation pressure threshold $p_{cav} = 4.86 \cdot 10^{-2}$ at the studied value of the dimensionless roughness was 0.15. Here we present the calculations for the case of unequal asperity heights and wavelengths on the surfaces and unequal wave numbers in the coordinate directions of each surface, the equal case being a particular one of the general solution.

As can be seen from eq. [3.61], the hydrodynamic pressure is governed by six parameters – the scaling factor, asperity height and wave number (both for the upper surface), asperity ratio, intra- and inter-surface wave ratios. The scaling factor is taken constant and equal to 1, which means equal evaluation lengths in the X- and Z-directions. Further, the dimensionless pressure is related to the relative gap and through it to the asperity height on the lower surface, so that varying the latter parameter affects the dimensionless cavitation pressure threshold; besides, variation of the asperity height on the lower surface entails that

Table 3.3 Reference values used in calculations

Parameter	Notation	Reference value	Dimensionality
Sliding velocity	u	8.9	m/s
Fluid dynamic viscosity	μ	3	cPa s
Clearance	Δ	6.4	μm
Control cell evaluation length	L	0.2	mm
Asperity height on lower surface	R_{a_1}	3	μm
Relative gap	$\Delta+2R_a\ h_0=1$	12.4	μm
Inter-surface wave ratio	ℓ	1/3 . . . 2	dimensionless
Wave number on upper surface, X-direction	k_{2_x}	1 . . . 9	units
Intra-surface wave ratio, upper surface	ℓ_k	1/2 . . . 3	dimensionless

of its upper counterpart at the same relative gap. To avoid this complication, all calculations were based on a constant asperity height A_1. By this means the number of factors to be studied was reduced from six to four.

Now the pattern of maximal pressures at different values of the mentioned parameters can be examined. By this means, we can find the excess of the pressure over the cavitation threshold and check for the presence of a positive hydrodynamic effect.

Figure 3.18 shows the dependence of the maximal pressure on the asperity height ratio A_1/A_2 in the range 2/3–2 for asperity heights $A_2=0.1, 0.15, 0.2$; inter-surface wave ratios $\ell=3, 1, 0.5, 0.33$; $A_1=0.15$ and $k_1=3$. These dependences are presented for intra-surface wave ratios $\ell_k=k_{1_z}/k_{1_x}=1/3$, 1, 2–Fig. 3.18(a), (b) and (c) successively. As can be seen, for

Figure 3.18 **Maximal pressures versus asperity height ratio A_1/A_2 at different intra-surface wave numbers k_{1_x}; asperity height of lower surface 0.15; and intra-surface ratios: (a) 2/3, (b) 1, (c) 2**

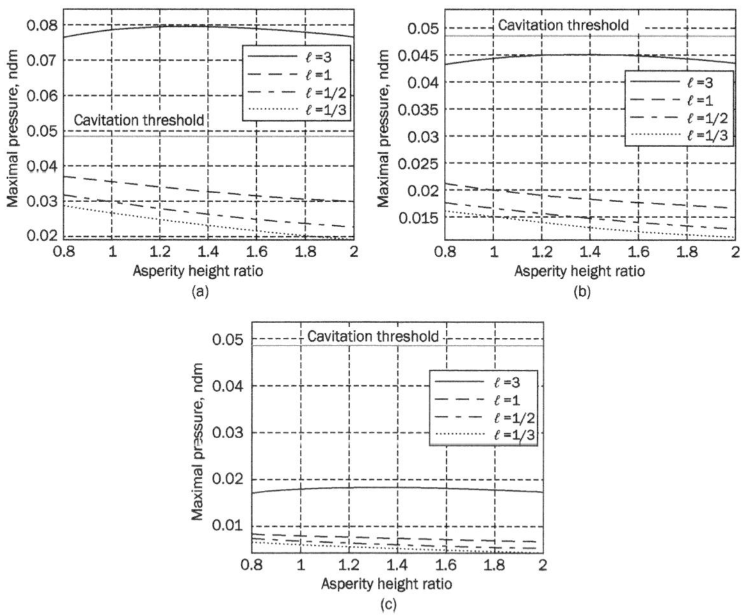

the wave ratio 3, the maximal positive pressure goes through a very weak maximum at asperity ratio above 1.3, but as a whole it decreases as the asperity ratio increases. Reduction of the asperity height on the upper surface reduces the maximal pressure at all studied values of the intra ratio. The different intra ratios evidence that when the wave number in the *Z*-direction is greater than in the *X*-direction (Fig. 3.18(a)) the maximal pressure is higher than in the opposite case, (Fig. 3.18(c)) or in that of equal wave numbers (Fig. 3.18(b)). This also demonstrates that inclusion of the second dimension strongly influences the calculated values of the hydrodynamic pressure, yielding a more reliable result.

Further, as can be seen from Fig. 3.18, the positive hydrodynamic effect is observed only in the case of intra ratio

$\ell_k = 1/3$ and inter ratio 3; as a whole, absence of the positive effect in the studied triangular version confirms that in reality the roughnesses are not a positive factor in lubrication.

Figure 3.19 presents the dependences of the maximal pressure for inter ratios in the range $\ell = 1/3$–4, at three different intra ratios $\ell_k = 1/3$, 1 and 2; it is actually a reconstruction of Fig. 3.18 with additional data for the lower surface wave number $k_{1_x} = 6$. As can be seen, beginning with wave ratio 2 the maximal pressures are lower than the cavitation threshold, so that the positive effect is absent at $\ell_k > 1$ for all studied roughness parameters.

The interesting fact here is that the $P_{max}(\ell)$ curves pass through an inflection point where asperity height change does not influence the maximal pressure, while beyond

Figure 3.19 Maximal and cavitation pressures versus inter-surface wave ratio at lower surface wave numbers $k_{1_x} = 3$ and 6, asperity height $A_1 = 0.15$; for intra ratios: (a) 2/3, (b) 1, (c) 2

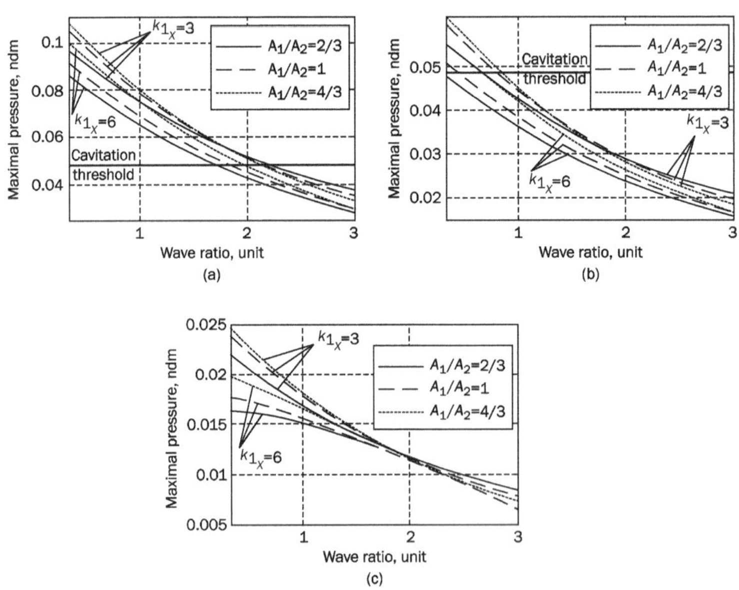

this point a larger asperity height leads to lower maximal pressure as should be expected – larger asperities increase lubrication resistance to the relative motion of the surfaces. Before the inflection point the asperity height on the upper surface is small (asperity ratios 2/3 and 1) and so is the number of asperities there (in our calculations k_{2_x} varies in the range 1–6 at $k_{1_x} = 3$), which makes for higher load support at smaller asperity heights. With larger wave numbers on the lower surface, $k_{1_x} = 6$, the inflection point apparently jumps to larger wave ratios; as a whole, as higher pressures are obtained at smaller wave ratios, it is preferable to produce surfaces with fewer waves and small amplitudes – which is unrealistic.

Comparison results for sinusoidal and triangular asperities (in the case of equal inter- and intra-roughness parameters $A_1/A_2 = 1$, $\ell_{k_1} = \ell_{k_2} = 1$) shows, Fig. 3.20, that the triangulars make for a maximal pressure about 25 per cent less than the sinusoidals in the range of small wave numbers; at larger wave numbers the discrepancies decrease up to less than 1 per cent. Comparison at $A_1 \neq A_2$ shows two pressures lower (by one-half and less) for the triangulars in the wave number range 2–19 with $A_1/A_2 = 2$; the maximal pressures were higher than the cavitation threshold at wave numbers between 1 to 3 only. All that is additional evidence of the detrimental role of roughness in fluid film hydrodynamics – seeing that the triangular configuration is the more realistic model.

All calculations for triangular roughness were carried out with the aid of a specially written Matlab program covering different input values of the roughness, roughness ratio, and intra- and inter-surface wave ratios. The program operated with sparse matrices with a view to economy in recourse to memory, and increased computing capacity at the larger wave ratios. Summarizing the reported observations for triangular roughness:

Figure 3.20 Cavitation threshold and maximal hydrodynamic pressure versus wave number; $\ell_{k_1} = \ell_{k_2} = 1$, $A_1 = A_2 = 0.15$ for triangular, and 0.25 for sinusoidal roughnesses

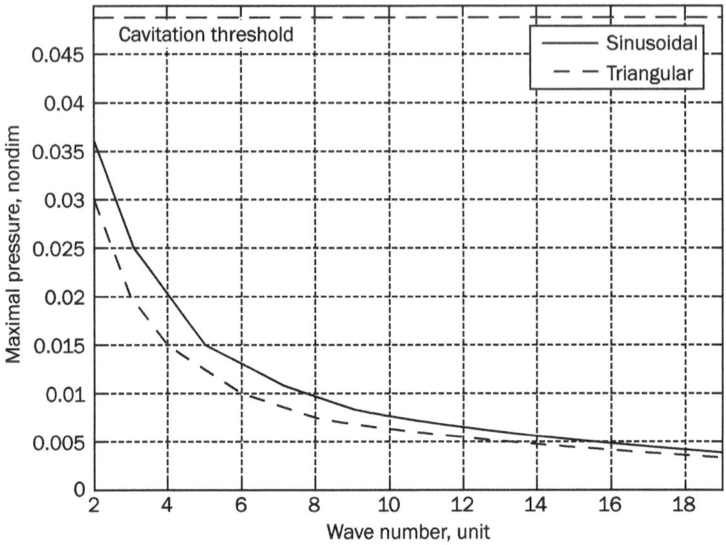

- The maximal support effect is reached at upper surface wave numbers three times those of the lower.
- The dependence of the maximal pressure on the surface wave number has an inflection point before which the maximal pressure is higher for lower wave ratios at smaller asperity heights.
- For larger asperity heights the load support decreases and the maximal hydrodynamic pressure is below the cavitation threshold.
- Above the intra surface wave ratio 2, the wave number in the *Z*-direction is double that in the *X*-direction, all pressures are below the cavitation threshold, so that no positive hydrodynamic effect is possible here.

3.7 Nomenclature

A – dimensionless asperity height of surface roughness (amplitude)

H, h – dimensionless and dimensional film thicknesses

P, p – dimensionless and dimensional hydrodynamic pressure

L – dimensional slider length/width

B, **C**, **D** – matrices for searching constants in pressure distribution equation

ε, e – dimensionless and dimensional journal eccentricity

c – dimensional radial clearance

C_1, C_2 – first and second constant of integration

k – number of waves on surface along coordinate; also, wedge coefficient in Section 3.5.1

K – vector in matrix used for finding constant of integration, triangular asperities

ℓ – wave ratio

ℓ_b – bearing scale factor

ℓ_s – inter-surface scale factor

L_x, L_z – dimensional surface length and width

M_x, M_z – dimensional mass flow rates per unit length

R – dimensional journal radii

R_a – dimensional asperity height (amplitude)

S – dimensionless local roughness step

V_x, V_y, V_z – fluid film velocities along x, y, z directions

T, t – dimensionless and dimensional time

u, v – dimensional velocitiy of sliding surface, along coordinate x and y respectively

(X, Y, Z), (x, y, z) – dimensionless and dimensional Cartesian coordinates

w – dimensional normal load, otherwise dimensional fluid film carrying capacity

Δ – dimensional minimal film thickness/gap between surfaces

δ_A, δ_{A1}, δ_{A2} – gap line unitary functions

λ – wavelength of surface profile along coordinate dimension, dimensional
μ – dimensional dynamic viscosity of fluid
θ – angular coordinate (bearings) or combined coordinate (roughness)
ν – dimensional kinematic viscosity of fluid
ω – angular velocity
τ – dimensional fluid film shear stresses near surface
f – dimensional forces acting on a fluid film
f_f – friction force
Φ, φ – dimensionless and dimensional phase shift (displacement) of sinusoidal roughness of upper surface relative to that of lower

3.8 Subscripts

0 – relative/initial value
1 and 2 – moving(upper) and stationary (lower) surface, respectively
A or a – amplitude
b – bearing
c – constant
cav – cavitation
e – extreme point of gap profile
f – friction
i – extreme point number
max – maximal value
t – time-dependent value
x, z – coordinate directions

Superscript (i) is identical with subscript i.

3.9 Acknowledgement

I wish to thank Mr. E. Goldberg, former resident scientific editor of the Technion, for valuable help in editing the chapter.

3.10 References

Anno, J. N., Walowit, J. A. and Allen, C. M. (1968), 'Microasperity lubrication', *ASME Journal of Lubrication Technology* **91**(2): 351–5.

Anno, J. N., Walowit, J. A. and Allen, C. M. (1969), 'Load support and leakage from microasperity – lubricated face seals', *ASME Journal of Lubrication Technology* **92**(4): 726–31.

ANSI/ASME Standard B46.1–1985 (1985) 'Surface Texture', New York, American Society of Mechanical Engineers.

Arghir, M., Roucou, N., Helene, M. and Frene, J. (2003), 'Theoretical analysis of the incompressible laminar flow in a macro-roughness cell', *ASME Journal of Tribology* **125**(2): 309–18.

Bernard, J. H., Steven, R. S. and Bo, O. J. (2004), *Fundamentals of Fluid Film Lubrication*, second edition, Marcel Dekker Inc.

Booker, J. F. (1965), 'A table of the journal-bearing integrals', *J Basic Eng, Transactions of the ASME* **87**(2): 533–5.

Booser, E. R. (1988), *CRC Handbook of Lubrication: Theory and Design*, 2, CRC Press, Boca Raton.

Bronshtein, I. N. and Semendyayev, K. A. (2007), *Handbook of Mathematics*, 5th edition, Springer, 1164.

Burstein, L. (2006), 'Two-sided surface roughness and hydrodynamic pressure distribution in lubricating films', *Lubrication Science* **19**(2): 101–12.

Burstein, L. (2008), 'Effect of sinusoidal roughened surfaces on pressure in lubricating film', *Int J Surface Science and Engineering* **2**(1–2): 52–70.

Burstein, L. (2009), 'Sinusoidal roughness in hydrodynamic lubrication. New transient 2D solution', in Davim, J. P., *Tribology Research Advances*, New York, Nova Publishers.

Burstein, L. (2010), 'Hydrodynamic lubrication of roughened surfaces with different asperity heights and wave numbers', *Journal of Tribology and Surface Engineering* **1** (in print).

Burstein, L. and Ingman, D. (1999), 'Effect of pore ensemble statistics on load support of mechanical seals with pore-covered faces', *ASME Transactions, Journal of Tribology* **121**(44): 927–32.

Burstein, L. and Ingman, D. (2000), 'Pore ensemble statistics in application to lubrication under reciprocating motion', *Tribology Transactions* **43**(2): 205–12.

Dowson, D., Taylor, L. M., Godet, M. and Berthe, D. (1978), 'Surface Roughness Effect in Lubrication. Leeds–Lyon Symposium on Tribology' (4th Symposium, 1977, Lyon, France), London, Mechanical Engineering Publications.

Labiau, A., Ville, F., Sainsot, P., Querlioz, E. and Lubrecht, T. (2008), 'Effect of sinusoidal surface roughness under starved conditions on rolling contact fatigue', *Proceedings of the IMechE, Part J: Journal of Engineering Tribology* **222**(3): 193–200.

Lai, T. W. (1994), 'Development of non-contacting spiral groove liquid face seals', *Lubrication Engineering* **50**(8): 625–40.

Letallear, N., Plouraboue, F. and Prat, M. (2002), 'Average flow model of rough surface lubrication: flow factors for sinusoidal surfaces', *ASME Transactions, Journal of Tribology* **124**(3): 539–46.

Olver, A. V. and Dini, D. (2007), 'Roughness in lubricated rolling contact: the dry contact limit', *Proc. IMechE, Part J: J. Engineering Tribology* **221**(7): 787–91.

Patir, N. and Cheng, H. S. (1978), 'An average flow model for determining effects of three-dimensional roughness on partial hydrodynamic lubrication', *ASME Journal of Lubrication Technology* **100**: 12–17.

Patir, N. and Cheng, H. S. (1979), 'Application of average flow model to lubrication between rough sliding surfaces', *ASME Journal of Lubrication Technology* **110**: 220–30.

Petrov, N. P. (1883), 'Friction in machines and the effect of the lubricant', *Inzh. Zh. St. Petersburg* **1**: 71–140, 227–9; **3**: 377–436; **4**: 535–64.

Polyanin, A. D. and Manzhirov, A. V. (2006), *Handbook of Mathematics for Engineers and Scientists*, London, Boca Raton, Chapman & Hall/CRC Press.

Reynolds, O. (1886), 'On the theory of lubrication and its application to Mr. Beauchamp Tower's experiments, including on experimental determination of the viscosity of olive oil', *Philos Trans R Soc* **177**: 157–234.

SRM 2073a (1995), 'Sinusoidal roughness specimens', *National Institute of Standards and Technology*, Springfield, VA.

Sommerfeld, O. (1904), 'Zur hydrodynamicshen Theorie der Schmiermittelreibung', *Z Angw Math Phys* **50**: 97–155.

Tønder, K. (1999), 'Vibration in sprig-supported bearing pads due to non-contacting roughnesses', *Wear* **232**(2): 256–61.

Tønder, K. (2004), 'Hydrodynamic effects of tailored inlet roughnesses: extended theory', *Tribology International* **37**(2): 137–42.

Totten, G. E., Westbrook, S. R. and Shah, R. J. (2003), *Fuels and Lubricants Handbook: Technology, Properties, Performance, and Testing*, ASTM.

Tower, B. (1885), 'Second Report on Friction Experiments (Experiments on Oil Pressure in a Bearing)', *Proc Inst Mech Eng*, 58–70.

Yang, D. and Liu, Y. (2008), 'Numerical simulation of electroosmotic flow in microchannels with sinusoidal roughness', *Colloids and Surfaces A: Physicochem. Eng. Aspects* 328(1–3): 28–33.

4

Micro/nano tribology

K. Mylvaganam and L. C. Zhang, University of New South Wales, Australia

Abstract: This chapter discusses the micro/nano sliding of materials in terms of adhesion, wear, friction and lubrication, using both experimental and theoretical methods such as SFA, STM, AFM, FFM and molecular dynamics simulation. A focus is to understand some characteristic phenomena associated with micro/nano tribology such as the scale effect of friction and wear deformation transition on no-wear, adhesion, ploughing and cutting regimes.

Keywords: adhesion, wear, friction, lubrication, molecular dynamics.

4.1 Introduction

Micro/nano tribology concerns the friction and wear of two objects in relative sliding whose dimensions range from micro-scales down to molecular and atomic scales. Unlike macro-tribology in which wear is inevitable, in micro/nano tribology wear is very light and the properties of contact surfaces often dominate the tribological performance.

It has been found that the adhesion force of a micro-scale object is over a million times greater than the force of gravity. This is because the adhesion force decreases linearly with size, whereas the gravitational force decreases with the size cubed (Kendall, 1994). Thus, micro objects adhere to their neighbours or surfaces and this is an obstacle to the miniaturization of components. Reduction in adhesion and friction were realized by applying principles of surface chemistry and tribology to micro-electromechanical systems (MEMS) that have a characteristic length of 100 nm to 1 mm, and nano-electromechanical systems (NEMS) that have a characteristic length of less than 100 nm. Contemporary examples in which atomic friction and wear play a central role are the optimal design, fabrication and operation of devices with atomic resolution, such as micro-machines and high-density magnetic recording systems. Over the last two decades or so, many studies have been carried out to explore the mechanisms of nano-friction and nano-wear, both theoretically and experimentally.

In this chapter we will discuss the micro/nano tribological investigations using surface force apparatus (SFA), scanning tunnelling microscope (STM), and atomic force microscopy (AFM) as well as friction force microscopy (FFM) (a subsequent modification of AFM). We will then present an overview on the capability of various methods for theoretical investigations. The molecular dynamics modelling to characterize the nanodeformation mechanisms will be demonstrated using diamond-copper and diamond-silicon systems as examples.

4.2 Experimental investigation

The study of tribology at micron and nanometre scales has become experimentally possible with the invention of SFA,

STM, AFM and FFM. Although all these are small-scale techniques, their measurement capabilities are different. For example, the SFA measures the interaction between molecularly smooth surfaces separated by a thin lubricant film; with which the surface separation, area of contact, lateral forces, and normal forces can be simultaneously measured. However, the AFM brings a sharp tip into contact with a lubricated surface, in which the film thickness and exact area of contact are unknown, while measuring the lateral and normal forces. Furthermore, the magnitude of the applied load, size of the contact area, and the composition of the probe surfaces in the above techniques are different. Table 4.1 compares some features of the SFA, STM and AFM techniques.

4.2.1 SFA analysis

The SFA developed in 1968 (Tabor and Winterton, 1969) and improved by Israelachvili and Tabor (Israelachvili and Tabor, 1972) is a common tool in the study of both static and dynamic properties of molecularly thin films sandwiched

Table 4.1 **Comparison of some features of SFA, STM and AFM**

Parameters	SFA	STM	AFM
Applied load	1–200 mN	N/A	1–100 nN
Size of the contact area	~10^{-5} cm^2	N/A	~10^{-13} cm^2
Sliding velocity	0.5–5 μm/s	0.02–200 μm/s	0–200 μm/s
Probe	Mica (generally)	Tungsten	Si_3N_4/Si/ diamond
Substrate requirement	Atomically smooth	Electrically conducting	Any

between two molecularly smooth surfaces. In the SFA, which comprises crossed cylinders of atomically smooth cleaved mica, the sliding friction force, surface separation and the area of contact can be measured simultaneously for a range of loads. The principal tribological application has been to study boundary lubrication with films a few molecules thick, although several friction measurements with dry surfaces have also been carried out.

A direct measurement of van der Waals forces in air between sheets of mica was made by accurately measuring the separation using multiple beam interferometry with an accuracy of ±0.3 nm (Tabor and Winterton, 1969). In this experiment one surface was held on a rigid support and the other on a light cantilever beam. As the surfaces were brought together, at a certain separation the surfaces jump into contact ('flick' together) when the attractive force between the surfaces overcomes the stiffness of the spring. Thus the 'flick' distance depends on the stiffness of the cantilever and this in turn provides a direct measure of the surface forces. Tabor and Winterton showed that 'normal' van der Waals forces predominate for separations less than 10 nm and the 'retarded' forces operate for distances greater than 20 nm.

Homola and co-workers (Homola et al., 1990) used SFA for simultaneous measurements of both the normal and the frictional forces between two molecularly smooth surfaces, the exact molecular contact area of the surfaces, the surface profile during sliding, and the distance between the two surfaces. They studied the sliding of mica surfaces in (i) dry atmosphere, (ii) controlled vapour atmospheres (i.e. in air or N_2), and (iii) with the surfaces immersed in bulk liquids. The interaction forces associated with this interfacial sliding was found to be much more localized than in the case of normal friction or boundary lubrication where the two surfaces separated by thin layers of some lubricant in which plastic deformations and damage occur during sliding. The work showed that at low loads the frictional force is described by the equation originally proposed by Bowden and Tabor: $F=S_cA$, where A is the molecular contact area and S_c is the critical shear stress (Bowden and Tabor, 1967). It was found that the dependence of A on the load is well described by the JKR theory (for adhesive contacts) and the Hertz theory (for non-adhesive contacts) even during sliding. At higher loads, adhesion is destroyed, multiasperity contact is established and the frictional force (F) becomes proportional to the load (L), analogous to Amontons' law, $F=\mu L$. In the presence of water vapour the friction decreased but the area of contact did not change, showing that the adhesion was maintained.

4.2.2 STM studies

The pioneering work on surface topography obtained with STM was published in 1982 by Binning and co-workers. In STM a sharp tip was systematically scanned over a sample surface in order to obtain information on the tip-sample interaction down to the atomic scale. As it used the tunnelling current between the conducting tip and a conducting sample, thus named the 'scanning tunnelling microscope', STM can only be used to study surfaces which are electrically conductive to some degree.

Since its invention, STM has provided images of surfaces and absorbed atoms and molecules with unprecedented resolution (Binning et al., 1982, Becker et al., 1985). An experimental result on surface topography was obtained in 1982 in which the surface reconstructions for (1 1 0) surfaces of $CaIrSn_4$ and Au were shown. The very high resolution of the STM rests on the strong dependence of the tunnel current on the distance between the metal tip and the scanned surface. STM has also been used to modify surfaces by locally pinning molecules to a surface (Foster et al., 1988) and by the transfer of an atom from the STM tip to the surface (Eigler and Schweizer, 1990). The tip of an STM exerts a finite force, which contains both van der Waals and electrostatic contributions, on an atom. Hence the magnitude and the direction of this force may be tuned by adjusting the position and the voltage of the tip. In addition, generally less force is required to move an atom along the surface than to pull it away from the surface. Thus it is possible to set the parameters such that the tip can pull an atom across the surface while the atom remains bound to the surface. For example, Eigler and Schweizer positioned xenon atoms on a single-crystal nickel surface with atomic precision by first placing the tip above the xenon atom and then increasing the tip–atom interaction by changing the required tunnel current to a higher value, which caused the tip to move towards the atom. They then moved the tip across the surface to the desired destination, dragging the xenon atoms. The tunnel current was then reduced to terminate the attraction between xenon and the tip, leaving the xenon bound to the surface at the desired location.

4.2.3 AFM and FFM investigations

The atomic force microscope introduced in 1985 (Binnig et al., 1986) provided a method for measuring forces on a nano-scale between a probe tip and an engineering surface. The AFM typically uses sharp silicon nitride, silicon, or diamond probes, whereas no specific chemistry is required for the substrate surface. The tip is attached to a free end of a cantilever and is brought very close to the surface. The cantilever bends towards or away from the sample in response to attractive or repulsive forces and its deflection is detected by means of a laser beam which is proportional to the normal force applied to the tip. AFM can operate in contact mode and non-contact mode. In the contact mode, the tip makes soft physical contact of the sample and either scans at a constant height or under a constant force where the deflection of the cantilever is fixed; even atomic resolution images are obtained in this mode. In the non-contact mode, the tip operates in the attractive force region and the tip-sample interaction is minimized. This mode allows scanning and the cantilever of choice is the one having a high spring constant so that it does not stick to the sample surface at small amplitudes.

Unlike STM, AFM has been used for topographical measurements of surfaces on the nanoscale which may be either electrically conducting or insulating as well as measuring adhesion and electrostatic force. Localized deformation, tip-substrate interactions and environmental effects often make the results difficult to reproduce. In 1987, a research group (Mate et al., 1987) at IBM reported an observation where the atomic structure of a surface manifests itself directly in the dynamical frictional properties of an interface. In particular, when they slide a tungsten tip on the basal plane of graphite, they observed that the frictional force displayed atomic periodicity of the graphite surface. According to them the friction coefficient between a tungsten tip of radius 300 nm and a basal plane of a graphite grain was 0.012 at a normal load of 10 μN. Another early interesting study (Kaneko et al., 1988) on the sliding of a tungsten tip of radius 10 μm on a carbon sputtered surface measured frictional force of about 1 μN at zero normal force, indicating an infinite friction coefficient.

Subsequent modification of AFM led to the development of the Friction Force Microscope (FFM) which is used to measure forces in the scanning direction, i.e. the force of friction. An AFM/FFM tip sliding on a surface simulates just one contact whereas at most solid–solid interfaces contact occurs at many asperities. Even so, as these asperities are of different sizes and shapes, the effect of size on friction/adhesion can be studied using tips of different radii. Bhushan and his co-workers have used these instruments to study various tribological phenomena such as surface roughness, adhesion, friction, scratching, wear, detection of material transition, etc. (Bhushan and Ruan, 1994; Ruan and Bhushan, 1994; Bhushan, 1995; Koinkar and Bhushan, 1997; Zhao and Bhushan, 1998; Sundararajan and Bhushan, 2000; Bhushan and Sundararajan, 1998; Bhushan and Dandavate, 2000). For example, Ruan and Bhushan used FFM to study a freshly cleaved highly-oriented pyrolytic graphite (HOPG) surface and found that the atomic-scale friction and the topography exhibited the same periodicity but the peaks were displaced relative to each other, which was explained by the variation in interatomic force in the normal and lateral directions. In a different study (Ruan and Bhushan, 1994) the same authors have reported that the coefficient of nano-scale friction for a Si_3N_4 tip of radius 50 nm versus HOPG graphite, natural diamond and Si(100) are 0.006, 0.04 and 0.07 respectively compared to the macro scale values of 0.1, 0.2 and 0.4. Recently, Bhushan presented a comprehensive review of AFM/FFM studies on the significant aspects of nano-tribology, nano-mechanics and material characterization (Bhushan, 2005).

McGuiggan and co-workers compared the AFM measurements with SFA measurements and found that for two different thin polymer films, AFM results showed little difference in the friction whereas SFA results gave a large difference (McGuiggan et al., 2001). This was explained by the variations in size of the probe of AFM and SFA. The sharp AFM tip may penetrate the lubricant film leading to the tip contacting, at least partially, the substrate surface.

One then measures the friction between two solid surfaces and the lubricant has little effect. On the other hand, SFA with its large contact area (see Table 4.1) maintained a uniform film thickness between the probe and the surface. Hence in this case, differences in the friction measurements may be due to differences in the film thickness.

The above experimental investigations demonstrate that the tribological properties such as surface roughness, adhesion, friction, wear, detection of material transfer and boundary lubrication can be studied at the micro/nano scale. However, to understand any discrepancies emerging from the different experimental methods, and to properly characterize the deformation mechanisms during the micro/nano tribological processes, theoretical methods are often required.

4.3 Theoretical investigation

4.3.1 Introduction

The advent of super computers for large-scale atomic simulations has led to the development of computational micro- and nano-tribology. While quantum mechanics is ideal for very small models on the atomic scale and micro/continuum mechanics is powerful for analysing the objects of micro and macroscopic dimensions, molecular dynamics (MD) simulation provides a useful means for detailed characterization of materials on the nano scale. However, the small time steps used in these simulations tend to result in a sliding speed far exceeding that in physical reality. The advantage of using molecular dynamics lies in its capacity to handle relatively large molecular systems; this is hard to do using quantum mechanics. Hence the reliability of molecular dynamics in exploring atomistic deformation mechanisms such as phase transformations and dislocation emissions – to which micromechanics and continuum mechanics are not applicable. Thus, MD simulations provide useful insight into experimental observations that may lead to the discovery of new phenomena or act as drivers for new experiments that in turn may lead to an eventual solution to the engineering problems in tribology. Moreover, 3D-computer visualization and animation allow us to follow atomistic behaviour during simulation at different time steps that helps to understand the chemistry and mechanics of the processes.

The MD simulation of nano-deformation operations depends on a number of essential modelling factors such as the choice of atomic interaction potential, the generation of the initial molecular model of the material and its relaxation process, the control of simulation temperature, the selection of control volume size and the application of moving control volume technique, the determination of integration time steps, identification of a temperature conversion model, and the method of stress analysis. For a discussion on these important issues in studying the nano-tribology using molecular dynamics the readers may refer to Zhang and Tanaka (1998, 1999) and Cheong et al. (2001).

As stated in section 4.2.1, the importance of the atomic contact area to atomic friction is not difficult to understand if the JKR theory is recalled. This theory, while considering the effect of surface energy in its analysis, has implicitly indicated that the real contact area must be of great concern to sliding loads on the atomic scale. If looking into the details of contact sliding, we can have two primary situations (Zhang and Tanaka, 1998). When two surfaces are in sliding without foreign particles, they are in two-body contact sliding, as shown in Fig. 4.1(a). In this case, the interactions among surface asperities play a central role in the process of wear and friction. However, if some particles appear between

Figure 4.1 Schematic drawing of molecular dynamics modelling of the sliding processes: (a) two-body sliding, (b) three-body sliding

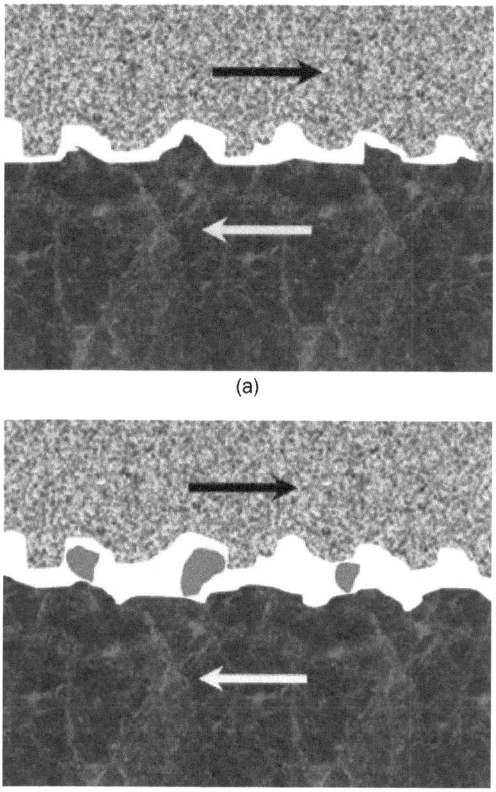

(a)

(b)

the surfaces, which could be the debris from worn surfaces or foreign particles due to contamination, a three-body contact sliding occurs, as shown in Fig. 4.1(b).

4.3.2 Diamond-copper sliding systems

Methods of modelling and analysis

For simplicity, an atomically smooth diamond asperity sliding on an atomically smooth surface of a copper monocrystal in its (1 1 1) plane has been considered. The variables of interest are the sliding speed *V*, indentation depth *d*, degree of surface lubrication or contamination and the tip radius of asperity *R* which is the radius of the envelope of centres of the surface atoms. The environmental temperature of the sliding system is 293 °K and the asperity rake angle is −60°. In addition, it is assumed that *d* keeps constant in a sliding process, which implies that the sliding system has infinite loop stiffness.

The initial model used in this sliding simulation consisted of a copper monocrystal work piece in its (1 1 1) plane with thermostat atoms arranged around the control volume to conduct the heat out properly and with boundary atoms arranged fixed to the space to eliminate the rigid body motion. A hemispherical diamond was used as the asperity. The interactions between copper–copper and copper–diamond atoms can be described by the modified Morse potential given by

$$\phi(r_{ij}) = \lambda_1 D\left[e^{-2\lambda_2 \alpha (r_{ij} - r_0)} - 2e^{\lambda_2 \alpha (r_{ij} - r_0)}\right], \qquad [4.1]$$

where r_{ij} is the interatomic separation between atoms *i* and *j* and r_0 is the equilibrium separation at which the potential is minimized. *D* and α are material constants listed in Table 4.2. With the above potential function available, the forces on atom *i* due to the interaction of all the other atoms can be calculated by

$$F_i = -\sum_{j=1, j \neq i}^{N} \Delta_i \phi\left(r_{ij}\right) \qquad [4.2]$$

where *N* is the total number of atoms in the model, (12000 <N<15000 is used in conjunction with the technique of moving control volume) including thermostat, boundary and diamond atoms. Consequently, the motion of all the Newtonian atoms in the control volume, including their instant position and velocity vectors, can be obtained by

Table 4.2 **Parameters in the standard Morse potential**

Parameter	C–Si	Cu–Cu	Cu–C
D (eV)	0.435	0.342	0.087
α (nm^{-1})	46.487	13.59	51.40
r_0 (nm)	0.19475	0.287	0.205
λ_1	1	1	(0,1)
λ_2	1	1	≥1

following the standard procedures of molecular dynamics analysis.

In principle, an asperity is three-dimensional and thus a three-dimensional molecular dynamics analysis would be more appropriate. However, a careful comparison showed (Tanaka and Zhang, 1996) that a two-dimensional model can lead to sufficiently accurate results in terms of the variations of temperature and sliding forces and easier characterization of deformation. We will therefore focus on the two-dimensional, plane-strain analysis in this section. When an instant configuration of the copper atomic lattice during sliding is obtained by the molecular dynamics analysis, the distribution of dislocations in the deformed lattice can be determined by the standard dislocation analysis (Courtney, 1990).

Mechanisms of wear

The deformation of the copper specimen has four distinct regimes under sliding, the no-wear regime, adhering regime, ploughing regime and cutting regime, as shown in Fig. 4.2. In the figure, the transition of deformation regimes is characterized by the non-dimensional indentation depth δ. When contact sliding takes place, δ is defined as d/R and can be viewed as a measure of the strain imposed by the diamond asperity.

Figure 4.2 **The transition of no-wear and wear regimes (the diamond slides from right to left)**

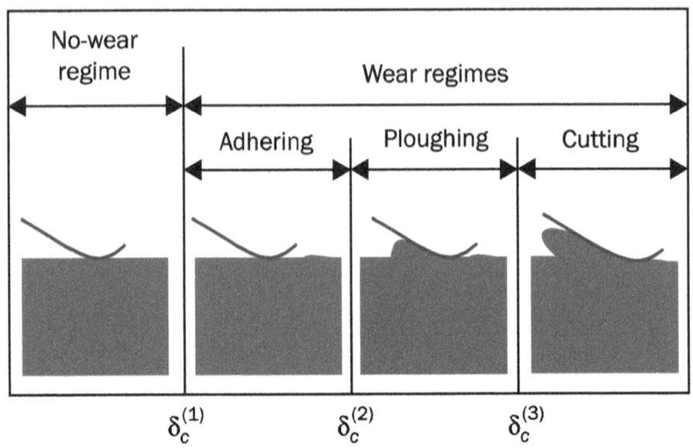

(Reprinted from Zhang and Tanaka (1997) with permission from Elsevier Science.)

In the no-wear regime, the atomic lattice of copper is deformed purely elastically. After the diamond asperity slides over, the deformed lattice recovers completely. In this case, sliding does not introduce any wear or initiate any dislocation.

When δ increases and reaches its first critical value, $\delta_c^{(1)}$, adhering occurs. The atomic bonds of some surface copper atoms are broken by the diamond sliding and these copper atoms then adhere to the asperity surface and move together with it. However, they may form new bonds with other surface atoms of copper and return to the atomic lattice. The above process repeats again and again during sliding, causing a structural change of the copper lattice near the surface, and creating surface roughness of the order of one to three atomic dimensions. In the meantime, some dislocations are also activated in the subsurface (Zhang and Tanaka, 1997).

If δ increases further to its second critical value, $\delta_c^{(2)}$, the above adhering deformation will be replaced by ploughing (Fig. 4.2). An apparent feature of deformation at this stage is that a triangular atom-cluster always exists in front of the leading edge of the diamond asperity and appears as a triangular wave being pushed forward. In this regime, the deformation zone in the subsurface becomes very large and a great number of dislocations are activated. MD simulation, by Zhang and Tanaka also showed the generation of grain boundaries by ploughing.

When δ reaches its third critical value, $\delta_c^{(3)}$, a new deformation state – cutting – appears, characterized by chip formation. Compared with the ploughing regime, the dimension of the deformation zone during cutting is smaller. Dislocations are distributed much more closely to the sliding interface.

Under some specific sliding conditions, not all the regimes would appear except the no-wear regime. For example, if the tip radius of the diamond asperity stays unchanged, but the sliding speed changes, then at lower sliding speeds all four regimes described above will appear. At higher speeds, however, the ploughing regime vanishes; see Fig. 4.3(a). On the other hand, at a given sliding speed, if the tip radius of the asperity is very small, say 1 nm, only the no-wear and

Figure 4.3 **Regime transition under specific sliding conditions: (a) non-dimensional indentation depth *vs* sliding speed, (b) non-dimensional indentation depth *vs* tip radius, (c) non-dimensional indentation depth *vs* lubrication/contamination**

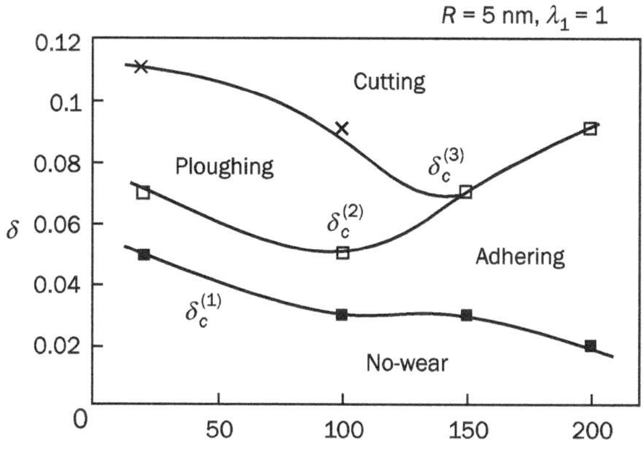

(a)

(*Continued*)

Figure 4.3 *Continued*

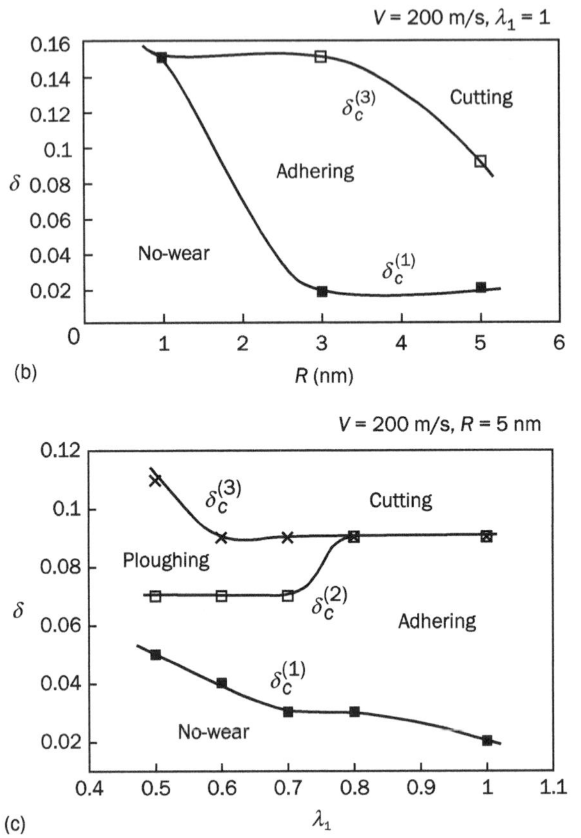

(Reprinted from Zhang and Tanaka (1997) with permission from Elsevier Science.)

cutting regimes emerge, as shown in Fig. 4.3(b). However, with relatively larger tip radii, adhering appears as a transition from no-wear to cutting. Another important factor that alters the deformation transition is the effect of surface lubrication or contamination. If the sliding interface is chemically clean, $\lambda_1 = \lambda_2 = 1$ in eq. [4.1]. In this case, as shown in Fig. 4.3(c), ploughing does not happen at a given sliding speed and tip radius. If the surface is lubricated, $\lambda_1 \leq 1$ with $\lambda_2 \geq 1$, and all the four regimes occur.

It is evident from Fig. 4.3 that the no-wear regime exists in a wide range of indentation depths. In addition, a smaller radius, a lower sliding speed, or better surface lubrication (i.e., smaller λ_1) enlarges the no-wear regime. This strongly indicates that the no-wear design of sliding systems may be realizable in practice. Moreover, it is important to note that the size of the no-wear regime is a strong function of sliding speed and surface lubrication. Therefore, sliding speed and lubrication should be taken into specific account in an attempt to design no-wear sliding systems. Recently, from a carbon-on-copper roller-sliding study, Jeng et al. (2005) reported that minimum resistance at the interface depends on the angular velocity of the roller and the separation distance between the roller and the slab. They found that a negative angular velocity minimizes wear and deformation at the interface.

The formation of various deformation regimes and their transition can be revealed by the variation of temperature distribution and dislocation motion in the atomic lattice. For instance, a larger indentation depth or a higher sliding speed indicates a higher input sliding energy, greater temperature rise and severe plastic deformation. This in turn means a higher density of dislocations with more complicated interactions in the deformed atomic lattice. For detailed information on the MD simulation results of temperature distribution, readers may refer to Zhang and Tanaka (1997).

Frictional behaviour

With the above deformation mechanisms in mind, the frictional behaviour of the system has been analysed. In the cutting regime, the variation of the conventional friction coefficient, $\mu = |F_x/F_y|$, is almost constant with the change of δ, where F_x and F_y are respectively the frictional force and normal indentation force during sliding. Thus it is clear that in this regime, F_x is proportional to F_y. In other regimes, however, the behaviour of F_x is complex. Particularly, μ becomes singular at a specific δ in the no-wear regime. The singularity of μ can be explained by examining the sliding forces when δ changes. On the atomic scale, the normal sliding force F_y always varies from attractive to repulsive. Thus at the transition point ($F_y = 0$), μ is infinite. The concept of the conventional coefficient of friction is no longer meaningful in the no-wear, adhering and ploughing regimes.

In non-contact sliding, the frictional force can be calculated by using eq. [4.2]. In contact sliding, Zhang and Tanaka obtained the following simple formula in terms of the contact area, based on their theoretical analysis:

$$F_x = \begin{cases} -\sum_{n=1}^{N_d}\sum_{m=1}^{N_c} \frac{\partial}{\partial x_n} \phi(r_{mn}), \text{ for } L_c = 0 \\ \zeta_1^{\mathrm{II}} L_c w_a + \zeta_2^{\mathrm{II}}, \text{ for } L_c^{(1)} < L_c \leq L_c^{(2)} \\ \zeta_1^{\mathrm{III}} L_c w_a + \zeta_2^{\mathrm{III}}, \textit{ for } L_c > L_c^{(2)} \end{cases} \qquad [4.3]$$

where $\zeta_1^{\mathrm{II}} = 409$ MPa, $\zeta_2^{\mathrm{II}} = 1.807 \times 10^{-8}$ nN, $\zeta_1^{\mathrm{III}} = 4.20$ GPa, $\zeta_2^{\mathrm{III}} = -1.899$ nN are constants, N_c is the total number of copper atoms in the model, N_d is the number of diamond atoms, L_c is the atomic contact length and $w_a = 0.226$ nm is the width of an atomic layer of copper in the direction perpendicular to its (1 1 1) plane. For $L_c = 0$, F_x is the resultant force of the atomic forces on all the diamond atoms in x-direction and can be derived directly from eq. [4.2]. For $L_c^{(1)} < L_c \leq L_c^{(2)}$ and $L_c > L_c^{(2)}$, the empirical expressions given in eq. [4.3] were obtained by fitting the MD simulation data in Fig. 4.4. The physical meaning of product $L_c w_a$ in eq. [4.3] is the atomic contact area in their sliding system. Equation [4.3] and Fig. 4.4, show that the frictional behaviour of an atomic sliding system cannot be described by a single formula.

Figure 4.4 **Relationship between the frictional force and contact length**

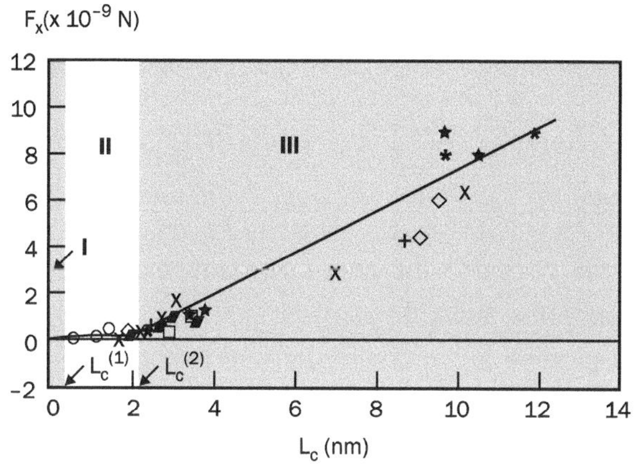

Notes

×: R=5 nm, V=20 m/s, λ_1=1;
∗: R=5 nm, V=100 m/s, λ_1=1;
★: R=5 nm, V=200 m/s, λ_1=1;
O: R=1 nm, V=200 m/s, λ_1=1;
◊: R=5 nm, V=200 m/s, λ_1=0.5;
■ R=5 nm, V=200 m/s, λ_1=0.6;
+: R=5 nm, V=200 m/s, λ_1=0.7;
□: R=5 nm, V=200 m/s, λ_1=0.8.

(Reprinted from Zhang and Tanaka (1997) with permission from Elsevier Science.)

There exist two distinct contact sliding zones, Zone II ($L_c^{(1)} < L_c \leq L_c^{(2)}$) and Zone III ($L_c > L_c^{(2)}$), where $L_c^{(2)} = 2.216$ nm is the transition boundary from Zones II to III, and $L_c^{(1)} = 0.277$ nm is the minimum contact length defined as the distance between two copper atoms in its (111) plane. The transition from non-contact to contact sliding is a sudden change because L_c does not exist below $L_c^{(1)}$. In Fig. 4.4, Zone II reflects the frictional behaviour of the system in the no-wear contact sliding, while Zone III shows it in the adhering and ploughing regimes. Thus $L_c^{(2)}$ can be interpreted physically as a critical contact length at which wear takes place. The data scattering in Fig. 4.4 indicates that other variables, such as the tip radius of asperity R and sliding speed V, also contribute greatly to friction. In other words, F_x should be a function of not only the contact area $L_c w_a$ but also V, R, and so on.

4.3.3 Diamond–silicon sliding systems

Modelling

Let us now consider the two-body and three-body contact sliding problems defined in Fig. 4.1. In the former, asperities are fixed on the sliding surfaces. To understand the fundamental deformation mechanism in a component induced by the penetration of asperities, researchers (Zhang and Tanaka, 1998; Mylvaganam and Zhang, 2009) developed a molecular dynamics model which consisted of a silicon monocrystal work piece in its (1 0 0) plane with thermostat atoms arranged around the control volume to conduct the heat out properly and with boundary atoms arranged fixed to the space to eliminate the rigid body motion. A hemispherical diamond was used as the asperity which should be irregular in reality, but it has been simplified in this study. Since diamond can be considered a rigid body compared with silicon, the model enables one to concentrate on the understanding of the deformation of silicon. In their simulation, Mylvaganam and Zhang used a large portion of the work material having 222,316 Si atoms as the control volume and performed the scratching with a diamond tip of radius 7.5 nm. In three-body contact sliding, the model shown in Zhang and Tanaka (1998) can be used, where the motion of a foreign particle between the two surfaces possesses both a translation and a self-rotation. To facilitate understanding, a single particle is considered for the time being and is approximated by a diamond ball of radius R, moving horizontally (translation) with a speed V_c and in the same time rotating about its centre independently with a peripheral speed V_r. When $V_r = 0$, the three-body contact sliding reduces to a two-body one. When $V_c = 0$ or $V_r = V_c$, on the other hand, it becomes a pure rolling process. The simulation model for the three-body contact problem used the technique of moving control volume.

Inelastic deformation

The molecular dynamics simulation showed that there always exists a thin layer of amorphous silicon in a specimen subsurface subjected to a two-body contact sliding, as shown in Fig. 4.5(a). This is in agreement with the experimental findings by Zhang and Zarudi (2001) (Fig. 4.5(b)). Gassilloud and co-workers observed nanocrystals embedded within the amorphous silicon when scratching at low speeds (Gassilloud et al., 2005). The thickness of the amorphous layer increases with increasing the penetration depth of asperity, δ. At some critical δ, dislocations can be developed in the crystalline silicon below the amorphous layer (Fig 4.5(c)). The fact that the amorphous layer appears for all δ during sliding shows that on the nanometre scale an inelastic deformation via amorphous phase transformation is a more energetically favourable mechanism. In the case with three-body contact sliding, the mechanism of inelastic deformation is the same, i.e., via amorphous phase transformation. However, because of the kinetic difference in the two-body and three-body sliding motions, the extent of subsurface damage is different. In general, a two-body contact sliding introduces a thinner amorphous layer. A three-body contact sliding, however, may yet leave a perfect crystalline structure after sliding, although wear has happened – this depends on the penetration depth of the particle, the ratio of its 'self-rotation' and 'translation' speeds and variation of atomic bonding strength affected by surface contamination. It was also found that the

Figure 4.5 **The subsurface microstructure of silicon monocrystals after a two-body contact sliding (the amorphous phase transformation has been predicted)**

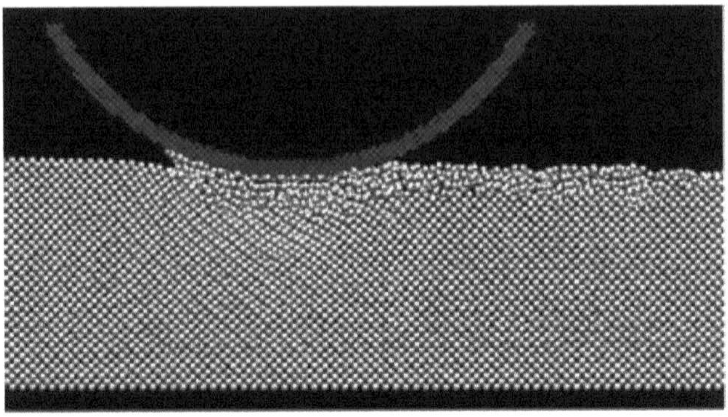

(a) A cross-sectional view of the deformed subsurface of the specimen (V_c=40 m/s, R=7.5 nm, δ=0.5 nm, sliding in [100] direction).

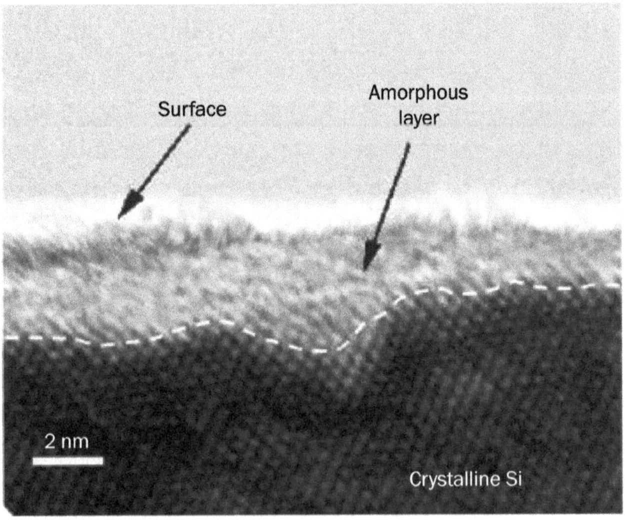

(b) An experimental result of the subsurface damage induced. Note the top amorphous layer (V_c=23.95 m/s, R=1 μm, δ=15.2 nm, sliding in [100] direction). Here, each spot represents a silicon atom.

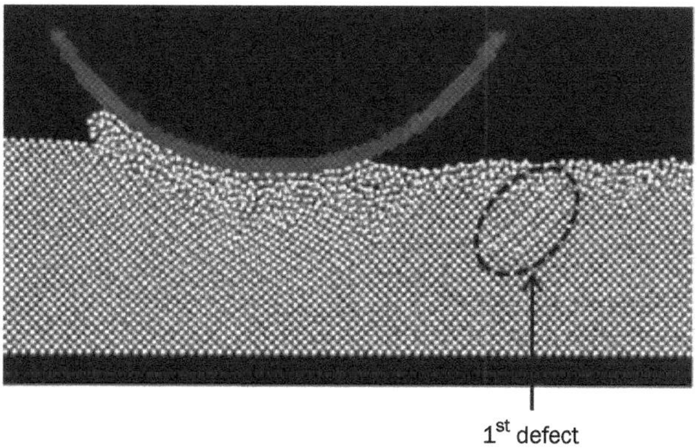

(c) A cross-sectional view of the deformed subsurface of the specimen (V_c=40 m/s, R=7.5 nm, δ=1.0 nm, sliding in [100] direction).

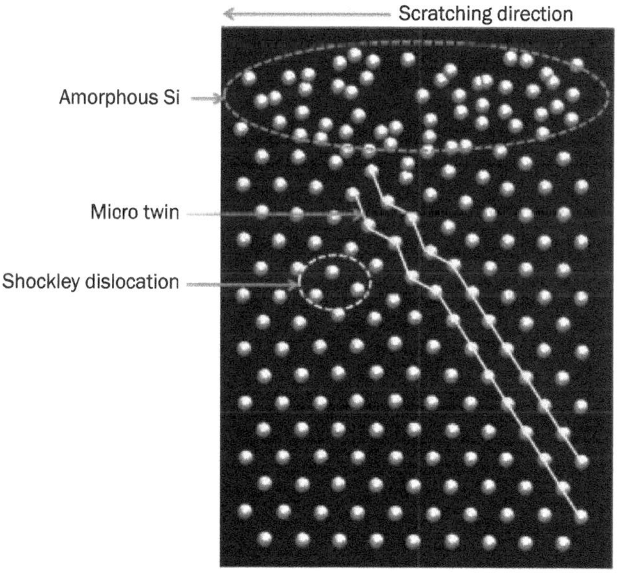

(d) Portion of the atoms on scratching when projected onto (1 1 0) plane showing nano-twin and Shockley dislocation.

(Reprinted from Zhang and Zarudi (2001) with permission from Elsevier Science.)

variation of sliding velocity from 20 m/s to 200 m/s does not change the deformation mechanisms described above.

Defect analysis

In the two-body contact problem, as the penetration depth increases, sliding with the large indenter generated a series of defects/dislocations in the silicon work piece (Fig. 4.5(c)). On analysing a one atomic layer thickness slice in the defect region shown in Fig. 4.5(c) by projecting the atoms on to (1 1 0) plane, showed the presence of nano-twins (Fig. 4.5(d)) with twin plane (1 1 1). The twinning effect stopped at the interior with Shockley partial dislocation located at the front boundaries of the nano-twins. In general, frequent twins and high density of uniform dislocations occur for materials with low stacking fault energy (SFE). Silicon has a low SFE of ~50 mJ/m^2. A very recent Transition Electron Microscopy (TEM) nanoscratch-induced deformation study of single crystal silicon also reported the nucleation of stacking faults and twins (Wu et al., 2009) although the tip radius and the load used in their experiment were much larger compared to those in MD simulation.

Wear regimes

Similar to the wear mechanisms for the diamond–copper sliding system (Zhang and Tanaka, 1997) discussed previously, the wear regimes of the current diamond–silicon system also depend on sliding conditions, as shown by the mechanism diagram Fig. 4.6. In a two-body contact sliding with a given sliding speed, the deformation of a silicon monocrystal falls into a no-wear, adhering, ploughing or cutting regime when the asperity penetration depth varies, as shown in the left half of the figure. Deformation without wear happens only under an extremely small penetration depth, when the atomic

Figure 4.6 **The wear diagram (diamond asperities/particles move from right to left; rotation of particles is anticlockwise)**

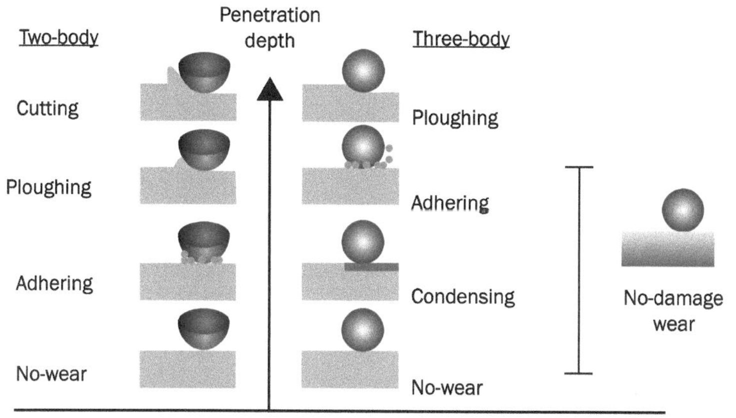

(Reprinted form Zhang and Tanaka (1998) with permission from Elsevier Science.)

lattice of silicon deforms purely elastically. With increasing penetration depth, adhering occurs, in which some surface atoms stick to the asperity surface and move together with it to cause wear. However, these atoms may return to the silicon substrate during sliding if the specimen surface has not been contaminated. When the penetration depth increases further, a new wear state, ploughing, characterized by an atomic cluster being pushed to move with the asperity, will appear. A further increase of the penetration depth leads to a continuous cutting process where the penetration depth of the asperity and the impingement direction has significant influence on the phase transformation and dislocation (Mylvaganam and Zhang, 2009).

In a three-body contact sliding, however, silicon will experience different wear regimes. They are the no-wear, condensing, adhering and ploughing regimes, as shown in the right half of Fig. 4.6. After the pure elastic deformation in the no-wear regime, the amorphous phase under the particle will experience a remarkable condensing locally without material removal. In other words, because the density of the surface silicon atoms under particle indentation becomes higher, condensing creates a sliding mark on the specimen surface. Thus condensing is a special wear process without material removal. A further particle penetration will lead to adhering and ploughing. These regimes are similar to the corresponding ones in the two-body contact sliding. Cutting rarely happens in three-body sliding processes but is possible if the particle penetration depth becomes sufficiently large and the self-rotation speed becomes small.

Another interesting phenomenon associated with the three-body contact sliding is the existence of a regime of no-damage wear. Under certain sliding conditions, the atomic bonding strength among surface silicon atoms can be weakened chemically. When this happens, these atoms can be removed via adhesion because diamond–silicon attraction is still strong. Due to the re-crystallization behind the particle, a worn specimen may appear as damage-free in the majority of its subsurface with only a little distortion within one or two surface atomic layers. In conjunction with the phenomenon that occurred in the condensing regime discussed above, it becomes obvious that a perfect subsurface after a three-body contact sliding does not necessarily indicate a no-wear process.

4.3.4 Effects of contact size and multiple asperities

Contact size

The investigation on diamond–copper sliding (Zhang and Tanaka, 1997), described in section 4.3.2, focused on the friction and wear mechanisms when the radius of the asperity

R is a constant while the depth of asperity indentation δ increases. Wear and plastic deformation consequently occur when δ reaches a critical value.

Based on some experimental observations (Carpick et al., 1996; Lantz et al., 1997), Hurtado and Kim (1999) proposed a micro-mechanical dislocation model of frictional slip, predicting that when the contact size is small the friction stress is constant and of the order of the theoretical shear strength. This is in agreement with AFM friction experiments. However, at a critical contact size there is a transition beyond which the frictional stress decreases with increasing contact size, until it reaches a second transition where the friction stress gradually becomes independent of the contact size. Hence, the mechanisms of slip are size-dependent, or in other words, there exists a scale effect. Before the first transition, the constant friction is associated with concurrent slip of the atoms without the aid of dislocation motion. The first transition corresponds to the minimum contact size at which a single dislocation loop is nucleated and sweeps through the whole contact interface, resulting in a single-dislocation-assisted slip. This mechanism is predicted to prevail for a wide range of contact sizes, from 10 nm to 10 μm, in radius for typical dry adhesive contacts. The second transition occurs for contact sizes larger than 10 μm, beyond which friction stress is once again constant due to cooperative glide of dislocations within dislocation pileups. The above dislocation model excludes wear or plastic deformation of the sliding parts.

To clarify this issue, Zhang et al. (2001) carried out a nano-tribology analysis using molecular dynamics by varying the asperity radius from 5 nm to 30 nm and keeping the indentation depth unchanged. The model consists of a single cylindrical asperity (rigid diamond) of various radii, sliding across a copper (1 1 1) plane with a speed of 5 m/s. The indentation depth, d, was 0.46 nm and −0.14 nm (0.14 nm above the work piece), respectively, where d is the distance between the surfaces of the asperity and specimen defined by the envelopes at the theoretical radii of their surface atoms. It must be noted that the molecular dynamics simulation cannot capture the second transition because it will require too long a computation time to analyse a model in the order of micrometres.

In the simulations where the asperity penetration depths are small enough so that there are no dislocations created within copper, the deformation corresponds to the no-wear regime described by Zhang and Tanaka (1997). In the case where the radius of the diamond asperity is less than 12 nm, the carbon atoms slide across the copper atoms in close contact. The surface of the copper work piece conforms closely to the shape of the asperity tip in contact (Fig. 4.7(a)). The variation of force with time steps showed strong indication of atomic stick-slip between the atoms of the asperity and the work piece. This implies that the sliding mechanism involved is similar to the ideal slip of two atomic planes in a perfect dislocation-free crystal. Hurtado and Kim (1999) referred to this sliding mechanism as concurrent slip.

When the asperity radius exceeds 12 nm, there are considerable differences in the sliding mechanism involved. The surface of the copper work piece does not conform closely to the shape of the carbon asperity (Fig. 4.7(b)) and the force variation shows little atomic stick-slip between the atoms of the asperity and the work piece. In addition, the friction stress is constant before the first transition but after which it decreases with the increasing contact width (Fig. 4.8). This figure further shows the indentation depth influences both the critical contact size at which first transition occurs and the rate of friction reduction after the transition. All these observations clearly indicate a change in the mechanism of sliding as predicted by Hurtado and

Figure 4.7 **Diamond asperity sliding on a monocrystalline copper surface**

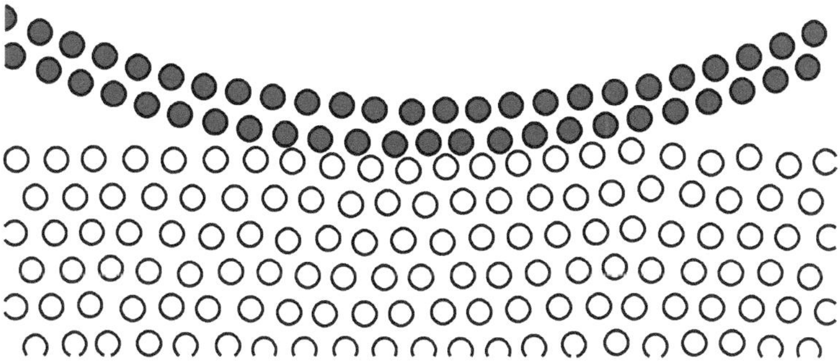

(a) Radius 8 nm – surface of copper workpiece conforms closely to shape of asperity with good contact.

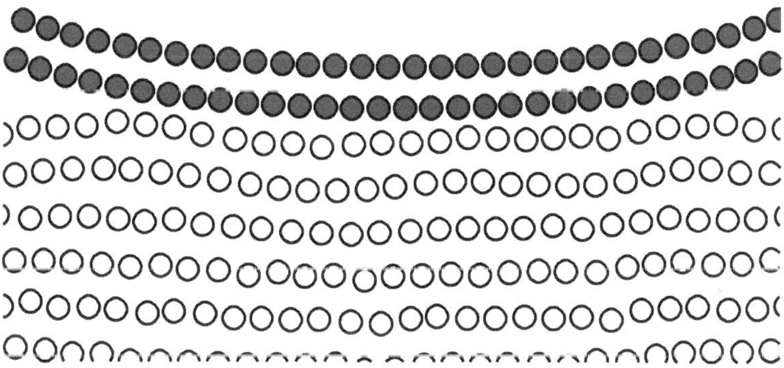

(b) Radius 30 nm – surface of copper workpiece does not conform closely to the shape of the asperity

(Reprinted from Zhang et al. (2001) with permission from Springer Publishers.)

Kim. However, at the greater indentation depth, permanent damage and wear occurred. Dislocation lines indicating plastic deformation 'within the body of the solid' are visible. This behaviour is similar to the adhering regime described by Zhang and Tanaka (1997).

Although in the MD simulation described above, only the first transition was observed, the fact that the friction stress decreases after this indicates that the coefficient of friction

Figure 4.8 **Frictional stress vs contact width for indentation depths of –0.14 nm and 0.46 nm**

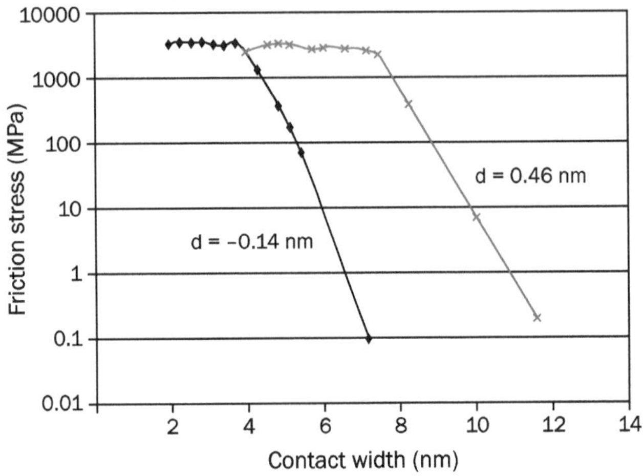

(Reprinted from Zhang et al. (2001) with permission from Springer Publishers.)

could change between the first and second transition. Thus in the experiment, depending on the probe radius and the load used, if the contact width falls within this region it is possible to get a different coefficient of nano-scale friction compared to macro-scale friction as observed in the latter part of section 4.2.3.

The contact width between the asperity and work piece obtained by the above molecular dynamics simulation can be compared with the predictions of the JKR theory (Johnson, 1985; Johnson et al., 1971) which shows that – for the present configuration of a circular cylinder in contact with a half space (plane-strain) – the indentation load per unit width on the asperity, P, and the contact width, 2a, follows the relationship

$$P = \frac{\pi E^* a^2}{4R} - \sqrt{2\pi E^* a w}, \qquad [4.4]$$

where R is the radius of the asperity, E^* is the effective modulus of the contact system (Johnson, 1985), w is the work of adhesion and can be determined by a nano-indentation simulation using molecular dynamics analysis. It is found that for the present diamond–copper (C–Cu) system, $w_{C\text{-}Cu} = 1.476\,J/m^2$. Since the diamond asperity is assumed to be a rigid body, the E^* in eq. [4.4] becomes 125.36 GPa by taking $E_C = \infty$, $E_{Cu} = 110$GPa and $\nu_{Cu} = 0.35$ (Callister, 1995).

Table 4.3 compares the contact widths from the molecular dynamics simulation, the JKR theory of eq. [4.4] and the Hertzian contact theory under various conditions. The values from the JKR and simulation are different, although the deformation of the copper work piece at $d = -0.14$ nm was purely elastic and that at $d = 0.46$ nm was almost purely inelastic. A possible cause is that the contact width of the molecular dynamics simulation contains the effect of sliding, while eq. [4.4] does not. It is also worth noting that compared to the predictions by the Hertzian contact theory, the

Table 4.3 **Contact lengths by the JKR and MD analyses for the case of diamond–copper interactions**

	Contact length 2a (nm)		
		$d=-0.14$ nm, $P=0.625$ N/m	$d=0.46$ nm, $P=22.969$ N/m
$R=5$ nm	JKR	2.914	3.764
	MD	2.870	4.120
	Hertz	0	2.160
		d=−0.14 nm, P=0.824 N/m	d=0.46 nm, P=27.34 N/m
$R=8$ nm	JKR	3.99	5.152
	MD	3.731	5.740
	Hertz	0	2.980

predictions by the JKR theory is much closer to the molecular dynamics results. This indicates that the effect of normal adhesion is considerable.

Sliding by multi-asperities

The single sliding asperity study has provided us with important knowledge on the deformation mechanisms of friction and wear of monocrystalline materials. In a real sliding system, however, a counterpart material is actually subjected to multi-asperity interactions, as illustrated in Fig. 4.1(a) and 4.1(b). When the first asperity has created a damaged zone, the material may deform differently under subsequent sliding interactions. Cheong and Zhang (2003) thus discussed the effects of the sliding by multi-asperities.

The mechanics model consists of three spherical diamond asperities, A, B and C, sliding on an atomically smooth silicon surface, as illustrated in Fig. 4.9. Their relative positions and orientations are defined by their distances, L_{AB} and L_{AC}, and angles with respect to the sliding direction, α and θ. There are three cases of special interest: (I) $\alpha = \theta = 0°$ with $L_{AB} < L_{AC}$, representing a repeated single-asperity sliding so that the effect of residual subsurface

Figure 4.9 The mechanics model for multi-asperity contact sliding

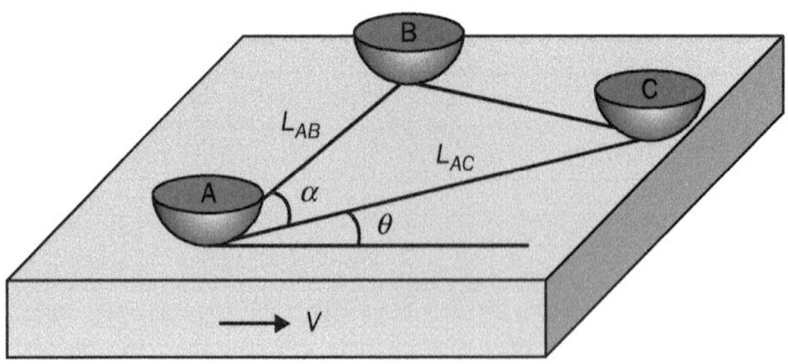

(Reprinted from Cheong and Zhang (2003) with permission from Interscience Publishers.)

damage can be understood; (II) $\alpha=0^{\circ}$ and $\theta=90^{\circ}$ with $L_{AB}=L_{AC}$, standing for the interaction of two parallel asperities; and (III) $\alpha=90^{\circ}$ and $\theta=0^{\circ}$, indicating the case with parallel sliding asperities coupled with an interaction from a third asperity. Again, since diamond is much harder than silicon, the asperities are modelled as rigid spheres. These spheres slide across the silicon surface at a specified velocity 40 m/s. The maximum depth of asperity penetration is 1.0 nm.

Case I

In this case, the second and third asperities B and C retrace the damaged path caused by asperity A, as in Fig. 4.10. Therefore, the cutting mechanism involved in the first and the following two asperities are very different. Asperity A cuts the silicon work piece in the same fashion as the case of a single sliding asperity, causing phase transformation of the original diamond cubic silicon. Asperity B, however, ploughs through the residual amorphous layer in the wake of asperity A. Again β-tin silicon forms beneath the asperity and then transforms into amorphous silicon when the asperity passes showing that the β-tin silicon phase is recoverable from the amorphous phase, provided that the required stress field is achieved. However, only some of the β-tin silicon is

Figure 4.10 **Cross-sectional view of silicon work piece and asperities A, B and C during sliding as in case I**

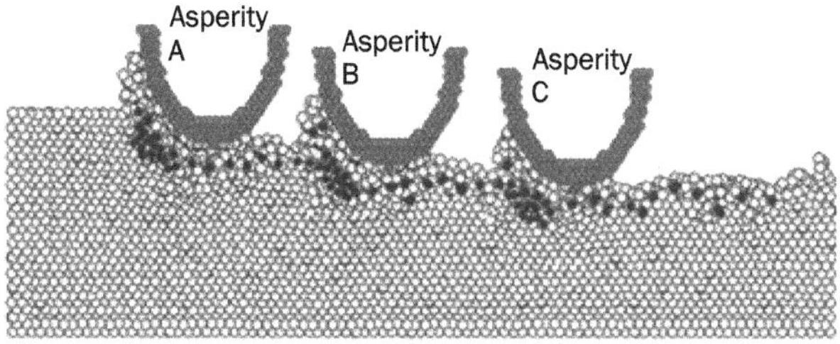

(Reprinted from Cheong and Zhang (2003) with permission from Interscience Publishers.)

recovered as asperities B and C retrace the amorphous damaged zone.

Cases II and III

In these cases, the asperities do not retrace the damaged zones. At the depth of asperity penetration of 1.0 nm, the wear mechanism observed is that of cutting and the plastic deformation due to the sliding asperities is very much localized. Figure 4.11 shows a cross-section of the silicon work piece through the centre of the asperities and it can be seen that there is almost no subsurface damage to the work piece between the two asperities. Localized plastic deformation occurs beneath the asperities.

The sliding asperities A, B and C also create trails of amorphous silicon within subsurface layers in the damaged zones, just as in nano-indentation (Cheong and Zhang, 2000). This is because, as the asperity slides across the silicon work piece, diamond cubic silicon continuously transforms into β-tin silicon beneath the asperity and then transforms into amorphous silicon when the asperity passes, leaving a layer of subsurface amorphous silicon in its wake.

Figure 4.11 Cross-section of the silicon workpiece through the centre of asperities (cases II and III)

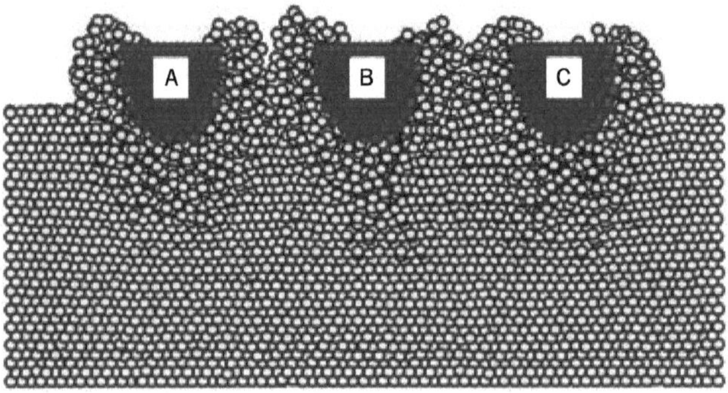

(Reprinted from Cheong and Zhang (2003) with permission from Interscience Publishers.)

Dislocations are absent at this particular depth of asperity penetration. This suggests that the plastic deformation is solely due to phase transformation.

4.4 Summary

This chapter has briefly discussed some fundamentals in studying the micro/nano-tribological properties such as adhesion, wear and friction using experimental techniques such as SFA, STM and AFM/FFM and the molecular dynamics technique. These techniques have different capabilities. In SFA the atomically smooth surface probe has a large contact area. This helps to have a film of uniform thickness between the two surfaces. In AFM, the sharp tip has a small contact area making it difficult to measure the film thickness and exact contact size, but it is capable of measuring the nano-tribological properties of any substrate.

The theoretical investigations using the MD technique have shown that in an atomic sliding system there generally exist four regimes, namely the no-wear regime, adhering regime, ploughing regime and cutting regime. The transition between different deformation regimes are governed by tip penetration depth, sliding speed, asperity geometry, and surface lubrication conditions. A smaller tip radius or a smaller sliding speed can bring about a greater no-wear regime. In addition, MD studies showed that a friction transition takes place at a critical contact size.

4.5 Note

1. Here, $12{,}000 \leq N \leq 15{,}000$ is used in conjunction with the technique of moving control volume.

4.6 References

Becker, R. S., Golovchenko, J. A., Hanmann, D. R. and Swartzentruber, B. S. (1985), 'Real-space observation of surface states on si(111) 7×7 with the tunnelling microscope', *Physical Review Letters* 55: 2032–4.

Bhushan, B. (1995), 'Micro/nanotribology and its applications to magnetic storage devices and mems', *Tribol. Int.* **28**: 85–95.

Bhushan, B. (2005), 'Nanotribology and nanomechanics', *Wear* **259**: 1507–31.

Bhushan, B. and Dandavate, C. (2000), 'Thin-film friction and adhesion studies using atomic force microscopy', *J. Appl. Phys.* **87**: 1201–10.

Bhushan, B. and Ruan, J. (1994), 'Atomic scale friction measurements using friction force microscopy. Part ii. Application to magnetic media', *ASME J. Trib.* **116**: 389–96.

Bhushan, B. and Sundararajan, S. (1998), 'Micro/nanoscale friction and wear mechanisms of thin films using atomic force and friction force microscopy', *Acta Mater.* **46**: 3793–804.

Binnig, G., Quate, C. F. and Gerber, C. (1986), 'Atomic force microscope', *Physical Review Letters* **56**: 930–3.

Binning, G., Rohrer, H., Gerber, C. and Weibel, E. (1982), 'Surface studies by scanning tunneling microscopy', *Physical Review Letters* **49**: 57–61.

Bowden, F. P. and Tabor, D. (1967), *Friction and Lubrication*, London: Methuen.

Callister, W. D. Jr. (1995), *Materials Science and Engineering – An Introduction*, New York: John Wiley & Sons.

Carpick, R. W., Agrait, D. F., Ogletree, D. F. and Salmeron, M. (1996), 'Variation of the interfacial shear strength and adhesion of a nanometer-sized contact', *Langmuir* **12**: 505.

Cheong, W. C. D. and Zhang, L. C. (2000), 'Molecular dynamics simulation of phase transformations in silicon monocrystals due to nano indentation', *Nanotechnology* **11**: 173.

Cheong, W. C. D. and Zhang, L. C. (2003), 'A stress criterion for the β-sn transformation in silicon under indentation and uniaxial compression', *Key Engineering Materials* 233–236, 603.

Cheong, W. C. D. and Zhang, L. C. (2003), 'Monocrystalline silicon subjected to multi-asperity sliding: nano-wear mechanisms, subsurface damage and effect of asperity interaction', *Int. J. Materials & Product Tech.* **18**: 398.

Cheong, W. C. D., Zhang, L. C. and Tanaka, H. (2001), 'Some essentials of simulating nano-surfacing processes using the molecular dynamics method', *Key Engineering Materials* **196**: 31.

Courtney, T. H. (1990), *Mechanical Behaviour of Materials*, Singapore: McGraw-Hill.

Eigler, D. M. and Schweizer, E. K. (1990), 'Positioning single atoms with a scanning tunneling microscope', *Nature* **344**: 524–6.

Foster, J. S., Frommer, J. E. and Arnett, P. C. (1988), 'Molecular manipulation using a tunnelling microscope', *Nature* **331**: 324–6.

Gassilloud, R., Ballif, C., Gasser, P., Buerki, G. and Michler, J. (2005), 'Deformation mechanisms of silicon during nanoscratching', *Applications and Materials Science* **202**: 2858–69.

Homola, A. M., Israelachvili, J. N., McGuiggan, P. M. and Gee, M. L. (1990), 'Fundamental experimental studies in tribology – the transition from interfacial friction of undamaged molecularly smooth surfaces to normal friction with wear', *Wear* **136**: 65–83.

Hurtado, J. A. and Kim, K-S. (1999), 'Scale effects in friction of single asperity contacts. Ii. Multiple-dislocation-cooperated slip', *Proc. R. Soc. London A* **455**: 3363.

Israelachvili, J. N. and Tabor, D. (1972), 'The measurement of van der waals dispersion forces in the range of 1.5 to 130 nm', *Proc. R. Soc. London A* **331**: 19–38.

Jeng, Y-R., Tsai, P-C. and Fang, T-H. (2005), 'Molecular dynamics studies of atomic-scale friction for roller-on-slab systems with different rolling-sliding conditions', *Nanotechnology* **16**: 1941–9.

Johnson, K. L. (1985), *Contact Mechanics*, Cambridge: Cambridge University Press.

Johnson, K. L., Kendall, K. R. and Roberts, A. D. (1971), 'Surface energy and the contact of elastic solids', *Proc. R. Soc. London A* **324**: 301.

Kaneko, R., Nonaka, K. and Yasuda, K. (1988), 'Scanning tunneling microscopy and atomic force microscopy for microtribology', *J. Vac. Sci. Tech.* **A6**: 291.

Kendall, K. (1994), 'Adhesion: molecules and mechanics', *Science* **263**: 1720–5.

Koinkar, V. N. and Bhushan, B. (1997), 'Effect of scan size and surface roughness on microscale friction measurements', *J. Appl. Phys.* **81**: 2472–9.

Lantz, M. A., O'Shea, S. L., Welland, M. E. and Johnson, K. L. (1997), 'Atomic-force-microscope study of contact area and friction on nbse$_2$', *Phys. Rev. B* **55**: 10776.

Mate, C. M., McClelland, G. M., Erlandsson, R. and Chiang, S. (1987), 'Atomic-scale friction of a tungsten tip on a graphite surface', *Physical Review Letters* **59**: 1942–5.

McGuiggan, P. M., Zhang, J. and Hsu, S. M. (2001), 'Comparison of friction measurements using the atomic force microscope and the surface forces apparatus: the issue of scale', *Tribol. Lett.* **10**: 217–23.

Mylvaganam, K. and Zhang, L. C. (2009), 'Nanoscratching-induced phase transformation of monocrystalline silicon – the depth-of-cut effect', *Adv. Mat. Res.* **76–78**: 387–91.

Ruan, J. and Bhushan, B. (1994), 'Atomic-scale and microscale friction of graphite and diamond using friction force microscopy', *J. Appl. Phys.* **76**: 5022–35.

Ruan, J. and Bhushan, B. (1994), 'Atomic-scale friction measurements using friction force microscopy: Part i general principles and new measurement technique', *ASME J Tribol.* **116**: 378–88.

Sundararajan, S. and Bhushan, B. (2000), 'Topography induced contributions to friction forces measured using an atomic force/friction force microscope', *J. Appl. Phys.* **88**: 4825–31.

Tabor, D. and Winterton, R. H. S. (1969), 'The direct measurement of normal and retarded van der waals forces', *Proc. R. Soc. Lond. A* **312**: 435–50.

Tanaka, H. and Zhang, L. C. (1996), in *Progress of Cutting and Grinding*, ed N Narutaki, Osaka: Japan Society for Precision Engineering, p. 262.

Wu, Y. Q., Huang, H., Zou, J. and Dell, J. M. (2009), 'Nanoscratch-induced deformation of single crystal silicon', *Journal of Vacuum Science & Technology B* **27**: 1374–7.

Zarudi, I., Cheong, W. C. D., Zou, J. and Zhang, L. C. (2004), 'Atomistic structure of monocrystalline silicon in surface nano-modification', *Nanotechnology* **15**: 104.

Zhang, L. C. and Tanaka, H. (1997), 'Towards a deeper understanding of wear and friction on the atomic scale-a molecular dynamics analysis', *Wear* **211**: 44–53.

Zhang, L. C. and Tanaka, H. (1998), 'Atomic scale deformation in silicon monocrystals induced by two-body and three-body contact sliding', *Tribol. Int.* **31**: 425–33.

Zhang, L. C. and Tanaka, H. (1999), 'On the mechanics and physics in the nano-indentation of silicon monocrystals', *JSME Int. J. Series A* **42**: 546–59.

Zhang, L. C. and Zarudi, I. (2001), 'Towards a deeper understanding of plastic deformation in mono-crystalline silicon', *Int. J. Mech. Sci.* **43**: 1985–96.

Zhang, L. C., Johnson, K. L. and Cheong, W. C. D. (2001), 'A molecular dynamics study of scale effects on the friction of single-asperity contacts', *Tribol. Lett.* **10**: 23.

Zhao, X. and Bhushan, B. (1998), 'Material removal mechanism of single-crystal silicon on nanoscale and at ultra-low loads', *Wear* **223**: 66–78.

5

Tribology in manufacturing

M. Jackson, Purdue University, USA and J. Morrell, Y12 National Security Complex, USA

Abstract: This chapter focuses on tribology in manufacturing processes from the viewpoint of understanding the fundamentals of sliding friction in those processes and the use of lubricants to control friction in manufacturing processes such as machining, drawing, rolling, extrusion, abrasive processes and processing at the micro and nanoscales. It is assumed that this chapter will serve as a focal point for engineers who are concerned with the role of tribology to maximize productivity and reduce costs associated with manufacturing processes under their command.

Keywords: tribology, manufacturing, lubrication, wear, friction.

5.1 Friction in manufacturing

The nature of the contact between surfaces is an important aspect of understanding the function of tribology in manufacturing processes, so macrocontacts and the stresses developed have been extensively studied and modelled by engineers. The properties of the materials in contact are homogeneous and isotropic. Macrocontact conditions are most useful in models for friction when there is lubrication and the effects of surface heterogeneities are of little importance.

Hertz's equations allow practitioners to calculate the maximum compressive contact stresses and contact dimensions for non-conforming bodies in elastic contact. The parameters required to calculate the quantities and the algebraic equations used for simple geometries are given in Table 5.1 (Young, 1989). It should be noted that Hertz's equations apply to static, or quasi-static, elastic cases. In the case of sliding, plastic deformation, contact of very rough surfaces, or significant fracture, both the distribution of stresses and the contact geometry will be altered. Hertz's contact equations have been used in a range of component design applications, and in friction and wear models in which the individual asperities are modelled as simple geometric contacts.

Greenwood and Williamson (1966) developed a surface geometry model that modelled contacts as being composed of a distribution of asperities. From that assumption, contact between such a surface and a smooth, rigid plane could be determined by three parameters: the asperity radius (R), the standard deviation of asperity heights (σ^*), and the number of asperities per unit area. To predict the extent of the plastic deformation of asperities, the plasticity index (ψ) (also a function of the hardness (H), elastic modulus (E), and Poisson's ratio (v)) was introduced

$$\psi = \left(\frac{E'}{H}\right)\left(\frac{\sigma^*}{R}\right)^{1/2} \qquad [5.1]$$

Table 5.1 **Equations for calculating elastic (Hertz) contact stress**

$$E^* = \left[\left(\frac{1-v_1^2}{E_1}\right)+\left(\frac{1-v_2^2}{E_2}\right)\right]$$

$$A = \left[\frac{(D_1+D_2)}{D_1D_2}\right] \qquad B = \left[\frac{(D_1-D_2)}{D_1\,D_2}\right]$$

Geometry	Contact dimension	Contact stress
Sphere-on-flat	$a=0.721\sqrt[3]{PDE^*}$	$S_c=0.918\sqrt[3]{P/(D^2E^*)^2}$
Cylinder-on-flat	$b=1.6\sqrt{pDE^*}$	$S_c=0.798\sqrt{p/DE^*}$
Cylinder-on-cylinder (axes parallel)	$b=1.6\sqrt{pE*/A}$	$S_c=0.798\sqrt{pA/E*}$
Sphere in a spherical socket	$a=0.721\sqrt[3]{PE*/B}$	$S_c=0.918\sqrt[3]{P(B/E*)^2}$
Cylinder in a circular groove	$b=1.6\sqrt{pE*/B}$	$S_c=0.798\sqrt{pB/E*}$

Key:
P – normal force;
p – normal force per unit contact length;
$E_{1,2}$ – modulus of elasticity for bodies 1 and 2, respectively;
$v_{1,2}$ – Poisson's ratios for bodies 1 and 2, respectively;
D – diameter of the curved body, if only one is curved;
$D_{1,2}$ – diameters of bodies 1 and 2, where $D_1 > D_2$ by convention;
S_c – maximum compressive stress;
a – radius of the elastic contact;
b – width of a contact (for cylinders);
$E*$ – composite modulus of bodies 1 and 2;
A, B – functions of the diameters of bodies 1 and 2.

where, $E' = E/(1-v^2)$. This basic formulation was refined by various investigators such as Whitehouse and Archard (1970) to incorporate other forms of height distributions, and the incorporation of a distribution of asperity radii, represented by the correlation distance β^*, which produced higher contact pressures and increased plastic flow. Therefore,

$$\psi = \left(\frac{E'}{H}\right)\left(\frac{\sigma^*}{\beta}\right)^{1/2} \qquad [5.2]$$

Hirst and Hollander (1974) used the plasticity index to develop diagrams to predict the start of scuffing wear. Other parameters, such as the average or root mean square slope of asperities, have been incorporated into wear models to account for such peculiarities (McCool, 1986). Worn surfaces are observed to be much more complex than simple arrangements of spheres, or spheres resting on flat planes, and Greenwood readily acknowledged some of the problems associated with simplifying assumptions about surface roughness (Greenwood, 1992). A comprehensive review of surface texture measurement methods have been given by Song and Vorburger (1992), while the most commonly used roughness parameters are listed in Table 5.2. Parameters such as skewness are useful for determining lubricant retention qualities of surfaces, since they reflect the presence of cavities. However, one parameter alone cannot precisely model the geometry of surfaces. It is possible to have the same average roughness (or RMS roughness) for two different surfaces.

Small amounts of wear can change the roughness of surfaces on the microscale and disrupt the nanoscale structure as well. Some of the following quantities have been used in models for friction:

- the true area of contact;
- the number of instantaneous contacts comprising the true area of contact;
- the typical shapes of contacts (under load);
- the arrangement of contacts within the nominal area of contact; and
- the time needed to create new points of contact.

Table 5.2 Definitions of surface roughness parameters

Let y_i = vertical distance from the ith point on the surface profile to the mean line N = number of points measured along the surface profile Thus, the following are defined:	
Arithmetic average roughness	$R_a = \frac{1}{N}\sum_{i=1}^{N}\lvert y_i \rvert$
Root-mean-square roughness	$R_q = \left[\frac{1}{N}\sum_{i=1}^{N} y_i^2\right]^{1/2}$
Skewness	$R_{sk} = \frac{1}{NR_q^3}\sum_{i=1}^{N} y_i^3$ A measure of the symmetry of the profile $R_{sk} = 0$ for a Gaussian height distribution
Kurtosis	$R_{kurtosis} = \frac{1}{NR_q^4}\sum_{i=1}^{N} y_i^4$ A measure of the sharpness of the profile $R_{kurtosis} = 3.0$ for a Gaussian height distribution $R_{kurtosis} < 3.0$ for a broad distribution of heights $R_{kurtosis} > 3.0$ for a sharply-peaked distribution

Finally, contact geometry-based models for friction generally assume that the normal load is constant. This assumption may be unjustified, especially when sliding speeds are relatively high, or when there are significant friction and vibration interactions in the tribosystem. As the sliding speed increases, frictional heating increases and surface thermal expansion can cause intermittent contact. The growth and excessive wear of intermittent contact points is termed thermoelastic instability (TEI) (Burton, 1980). TEI is only one potential source of the interfacial dynamics responsible for stimulating vibrations and normal force variations in sliding contacts. Another major cause is the eccentricity of rotating shafts, run-out, and the transmission of external vibrations. Static friction and stick-slip behaviour are considered, and as with kinetic friction, the causes for such phenomena can be interpreted on several scales.

5.1.1 Static friction and stick-slip

If all possible causes for friction are to be considered, it is reasonable to find out whether there are other means to cause bodies to stay together without the requirement for molecular bonding. Surfaces may adhere, but adherence is not identical to adhesion, because there is no requirement for molecular bonding. If a certain material is cast between two surfaces and, after penetrating and filling irregular voids in the two surfaces, solidifies to form a network of interlocking contacting points, there may be a strong mechanical joint produced, but no adhesion. Adhesion (i.e., electrostatically balanced attraction/chemical bonding) in friction theory meets the need for an explanation of how one body can transfer shear forces to another. Clearly, it is convenient to assume that molecular attraction is strong enough to allow the transfer of force between bodies, and in fact this assumption has led to many of the most widely used friction theories. From another perspective, is it not equally valid to consider that if one pushes two rough bodies together so that asperities penetrate, and then attempt to move those bodies tangentially, the atoms may approach each other closely enough to repel strongly, thus causing a backlash against the bulk materials and away from the interface. The repulsive force parallel to the sliding direction must be overcome to move the bodies tangentially, whether accommodation occurs by asperities climbing over one another, or by deforming one another. In the latter, it is repulsive forces and not adhesive bonding that produces sliding resistance. This section focuses on static friction and stick slip phenomena.

Ferrante et al. (1988) have provided a comprehensive review of the subject and a discussion of adhesion and its relationship to friction has been conducted by Buckley (1981). Atomic probe microscopes permit investigators to study adhesion and lateral forces between surfaces on the atomic scale. The force required to shift the two bodies tangentially must overcome bonds holding the surfaces together. In the case of dissimilar metals with a strong bonding preference, the shear strength of the interfacial bonds can exceed the shear strength of the weaker of the two metals, and the static friction force ($\mathbf{F}_s$) will depend on the shear strength of the weaker material (τ_m) and the area of contact (A). In terms of the static friction coefficient μ_s,

$$F_s = \mu_s \, P^* = \tau_m \, A \qquad [5.3]$$

or

$$\mu_s = (\tau_m / P^*) A \qquad [5.4]$$

where $\mathbf{P}^*$, the normal force, is comprised of the applied load and the adhesive contribution normal to the interface. Under specially controlled conditions, such as friction experiments with clean surfaces in vacuum, the static friction coefficients can be greater than 1.0, and the experiment becomes a test of the shear strength of the solid materials than of interfacial friction. Scientific understanding and approaches to modelling friction has been strongly influenced by concepts of solid surfaces and by the instruments available to study them. Atomic-force microscopes and scanning tunnelling microscopes permit views of surface atoms with high resolution and detail. Among the first to study nanocontact frictional phenomena were McClellan et al. (1987, 1988). A tungsten wire with a very fine tip is brought down to the surface of a highly oriented, cleaved basal plane of pyrolytic graphite as the specimen is oscillated at 10 Hz using a piezoelectric driver system. The cantilevred wire is calibrated so that its spring constant is known (2500 N/m) and the normal force could be determined by measuring the deflection of the tip using a reflected laser beam. As the normal force is decreased, the contributions of individual atoms to the tangential force became apparent. At the same time, it appeared that the motion of the tip became less uniform, exhibiting atomic-scale stick-slip.

Thompson and Robbins (1990) discussed the origins of nanocontact stick-slip when analysing the behaviour of molecularly thin fluid films trapped between flat surfaces of face-centred cubic solids. At that scale, stick-slip was believed to arise from the periodic phase transitions between ordered static and disordered kinetic states. Immediately adjacent to the surface of the solid, the fluid assumed a regular, crystalline structure, but this was disrupted during each slip event. The experimental data points of friction force per unit area versus time exhibited extremely uniform classical stick-slip appearance. Once slip occurred, all the kinetic energy must be converted into potential energy in the film. In subsequent papers, Robbins et al. (1991, 1993) used this argument to calculate that the critical velocity, v_c, below which the stick-slip occurs is:

$$v_c = c\left(\sigma \mathbf{F}_s / M\right)^{1/2} \qquad [5.5]$$

where σ is the lattice constant of the wall, $\mathbf{F}_s$ is the static friction force, M is the mass of the moving wall, and c is a constant.

Friction is defined as the resistance to relative motion between two contacting bodies parallel to a surface that separates them. Motion at the atomic scale is unsteady. In nanocontact, accounting for the tangential components of thermal vibrations of the atoms thus affects our ability to clearly define relative motion between surfaces. Under some conditions it may be possible to translate the surface laterally while the adhesive force between the probe tip and the opposite surface exceeds the externally applied tensile force. Landman et al. (1990) reviewed progress in the field of molecular dynamics (MD). By conducting MD simulations of nickel rubbing a flat gold surface, Landman illustrated how the tip can attract atoms from the surface simply by close approach without actual indentation. A connective neck or bridge of surface atoms was observed to form as the indenter was withdrawn. The neck can exert a force to counteract the withdrawal force on the tip, and the MD simulations clearly model transfer of material between opposing asperities under pristine surface conditions. Landman has subsequently conducted numerous other MD simulations, including complete indentation and indentation in the presence of organic species between the indenter and substrate. Belak and Stowers (1992), using a material volume containing 43,440 atoms in 160 layers, simulated many of the deformational features associated with metals, such as edge dislocations, plastic zones, and point defect generation. Calculated shear stresses for a triangular indenter passing along the surface exhibited erratic behaviour, not unlike that observed during metallic sliding under clean conditions. Pollock and Singer (1992) compiled a series of papers on atomic-scale approaches to friction.

While MD simulations and atomic-scale experiments continue to provide fascinating insights into frictional behaviour, under idealized conditions, most engineering tribosystems are non-uniform. Not only are surfaces not atomically flat, but the materials are not homogeneous, and surface films and contaminant particles of many kinds, much larger than the atomic scale, may influence interfacial behaviour. Static friction coefficients measured experimentally under ambient or contaminated conditions probably will not assume the values obtained in controlled environments. In a series of carefully conducted experiments on the role of adsorbed oxygen and chlorine on the shear strength of metallic junctions, Wheeler (1975) showed how μ_s can be reduced in the presence of adsorbed gases. On the other hand, static friction coefficients for pure, well-cleaned metal surfaces in the presence of non-reactive gases like He can be relatively high. It is interesting to note that the friction of copper on nickel and the friction of nickel on copper are quite different. This is not an error, but rather a demonstration of the fact that reversing the materials of the sliding specimen and the counterface surface can affect the measured friction, confirming the assertion that friction is a property of the tribosystem and not of the materials in contact. A cryotribometer was used to obtain the data in Table 5.3.

Table 5.3 Static friction coefficients for clean metals in helium gas at two temperatures

	Static friction coefficient	
Material combination	300 K	80 K
Fe (99.9%) on Fe (99.99%)	1.09	1.04
Al (99%) on Al (99%)	1.62	1.60
Cu (99.95%) on Cu (99.95%)	1.76	1.70
Ni (99.95%) on Ni (99.95%)	2.11	2.00
Au (99.98%) on Au (99.98%)	1.88	1.77
Ni (99.95%) on Cu (99.95%)	2.34	2.35
Cu (99.95%) on Ni (99.95%)	0.85	0.85
Au (99.98%) on Al (99%)	1.42	1.50
Fe (99.9%) on Cu (99.95%)	1.99	2.03

The length of time that two solids are in contact can also affect the relative role that adhesion plays in establishing the value of the static friction coefficient. Two distinct possibilities can occur: (a) if the contact becomes contaminated with a lower shear-strength species, the friction will decline; and (b) if the contact is clean and a more tenacious interfacial bond develops, the static friction will tend to increase. Akhmatov (1939) demonstrated that by using cleaved rock salt, the formation of surface films over time lowers static friction.

The opposite effect has been demonstrated for metals. A first approximation of rising static friction behaviour is given by:

$$\mu_{S(t)} = \mu_{S(t=\infty)} - \left[\mu_{S(t=\infty)} - \mu_{S(t=0)}\right] e^{-ut} \qquad [5.6]$$

where $\mu_s(t)$ is the current value of the static friction coefficient at time t, $\mu_s(t=\infty)$ is the limiting value of the static friction coefficient at long times, $\mu_s(t=0)$ is the initial static friction coefficient, and u is a rate constant. In contrast to exponential dependence on time, Buckley (1981) showed that by using data for tests of single-crystal Au touching Cu-5% Al alloy that junction growth can cause the adhesive force to increase linearly with time.

When materials are placed in intimate contact, it is not unexpected that the atoms on their surfaces will begin to interact. The degree of this interaction will depend on the contact pressure, temperature, and the degree of chemical reactivity that the species have for each other, hence, static friction can change with the duration of contact.

Despite the two opposite dependencies of static friction on time of contact, observations are consistent from a thermodynamic standpoint. Systems tend toward the lowest energetic state and in the case of interfaces, this state can be achieved either by forming bonds between the solids, or by forming bonds with other species (adsorbates and films) in the interface. The former process tends to strengthen the shear strength of the system, and the latter tends to weaken it. Sikorski (1964) reported the results of experiments designed to compare friction coefficients of metals with their coefficients of adhesion (defined as the ratio of the force needed to break the bond between two specimens to the force which initially compressed them together). Rabinowicz (1992) conducted a series of simple, tilting-plane tests with milligram- to kilogram-sized specimens of a variety of metals. Results demonstrated the static friction coefficient to increase as slider weight (normal force) decreased. For metal couples such as Au/Rh, Au/Au, Au/Pd, Ag/Ag, and Ag/Au, as the normal force increased over about six orders of magnitude (1 mg to 1 kg), the static friction coefficients tended to decrease by nearly one order of magnitude.

Under low contact pressures, surface chemistry effects can play a relatively large role in governing static friction behaviour. However, under more severe contact conditions, such as extreme pressures and high temperatures, other factors, more directly related to bulk properties of the solids, dominate static friction behaviour. When very high pressures and temperatures are applied to solid contacts, diffusion bonds or solid-state welds can form between solids, and the term static friction ceases to be applicable. Table 5.4 lists a series of reported static friction coefficients. Note that in certain cases, the table references list quite different values for these coefficients.

The temperature of sliding contact can affect the static friction coefficient. This behaviour was demonstrated for single crystal ceramics by Miyoshi and Buckley (1981), who conducted static friction tests of pure iron sliding on cleaned {0001} crystal surfaces of silicon carbide in a vacuum (10^{-8} Pa). For both <1010> and <1120> sliding directions, the static friction coefficients remained about level (0.4 and 0.5,

Table 5.4 **Static friction coefficients for metals and non-metals (dry or unlubricated conditions)**

Material combination			
Fixed specimen	**Moving specimen**	μ_s	**Table reference number**
Metals and alloys on various materials			
Aluminium	Aluminium	1.05	1
	Steel, mild	0.61	1
	Titanium	0.54	3
Al, 6061-T6	Al, 6061-T6	0.42	4
	Copper	0.28	4
	Steel, 1032	0.35	4
	Ti-6Al-4V	0.34	4
Brass	Steel, mild	0.53	1
Cast iron	Cast iron	1.10	1
Cadmium	Cadmium	0.79	3
	Iron	0.52	3
Chromium	Cobalt	0.41	3
	Chromium	0.46	3
Cobalt	Cobalt	0.56	3
	Chromium	0.41	3
Copper	Cast iron	1.05	1
	Chromium	0.46	3
	Cobalt	0.44	3
	Copper	1.6	2
	Glass	0.68	1
	Iron	0.50	3
	Nickel	0.49	3
	Zinc	0.56	3
Gold	Gold	2.8	2
	Silver	0.53	3
Iron	Cobalt	0.41	3
	Chromium	0.48	3
	Iron	0.51	3

(Continued)

Table 5.4 *Continued*

Material combination			
Fixed specimen	**Moving specimen**	μ_s	**Table reference number**
Iron	Tungsten	0.47	3
	Zinc	0.55	3
Indium	Indium	1.46	3
Lead	Cobalt	0.55	3
	Iron	0.54	3
	Lead	0.90	3
	Silver	0.73	3
Magnesium	Magnesium	0.60	1
Molybdenum	Iron	0.69	3
	Molybdenum	0.8	2
Nickel	Chromium	0.59	3
	Nickel	0.50	3
Niobium	Niobium	0.46	3
Platinum	Platinum	3.0	2
Silver	Copper	0.48	3
	Gold	0.53	3
	Iron	0.49	3
	Silver	1.5	2
Steel	Cast iron	0.4	2
Steel, hardened	Steel, hardened	0.78	1
	Babbitt	0.42, 0.70	1
	Graphite	0.21	1
Steel, mild	Steel, mild	0.74	1
	Lead	0.95	1
Steel, 1032	Aluminium	0.47	4
	Copper	0.32	4
	Steel, 1032	0.31	4
	Ti-6Al-4V	0.36	4

(Continued)

Table 5.4 *Continued*

Steel, stainless 304	Copper	0.33	4
Tin	Iron	0.55	3
	Tin	0.74	3
Titanium	Aluminium	0.54	3
	Titanium	0.55	3
Tungsten	Copper	0.41	3
	Iron	0.47	3
	Tungsten	0.51	3
Zinc	Cast iron	0.85	1
	Copper	0.56	3
	Iron	0.55	3
	Zinc	0.75	3
Zirconium	Zirconium	0.63	3

1. *Handbook of Tribology, B.* Bhushan and B. K. Gupta, McGraw Hill (1991).
2. *Handbook of Chemistry and Physics,* 48th edn., CRC Press (1967).
3. E. Rabinowicz, *ASLE Trans.,* Vol. 14, p. 198; plate sliding on inclined plate at 50% rel. humidity (1971).
4. 'Friction Data Guide', General Magnaplate Corp., Ventura, California 93003, TMI Model 98-5 Slip and Friction Tester, 200 grams load, ground specimens, 54% rel. humid., average of 5 tests (1988).

respectively) from room temperature up to about 400°C; then they each rose by about 50% as the temperature rose to 800°C. The authors attributed this effect to increased adhesion and plastic flow.

The role of adsorbed films on static friction suggests that one effective strategy for alleviating or reducing static friction is to introduce a lubricant or other surface treatment to impede the formation of adhesive bonds between mating surfaces. Contamination of surfaces from exposure to the ambient environment performs essentially the same function, but is usually less reproducible. Campbell (1940) demonstrated how the treatment of metallic surfaces by oxidation can reduce the static friction coefficient. Oxide films were produced by heating metals in air, while sulfide films were produced by immersing the metals in sodium sulfide solution. Except for the film on steel, film thicknesses were estimated to be 100–200 nm. Results from ten experiments, using a three ball-on-flat plate apparatus, were averaged to obtain static friction coefficients. In addition to producing oxides and sulfides, Campbell also tested oxide and sulfide films with Acto oil. The results of this investigation are shown in Table 5.5. For copper, the static friction coefficient ($\mu_s = 1.21$, with no film) decreased when the sulfide film thickness was increased from 0 to about 300 nm, after which the static friction coefficient remained about constant at 0.66.

The extent to which the solid lubricant can reduce static friction may be dependent on temperature, as confirmed by Hardy's earlier studies on the static friction of palmitic acid films on quartz. Between 20 and 50°C, the static friction coefficient decreases until melting occurs, at which time the lubricant loses its effectiveness.

Stick-slip is often referred to as a relaxation-oscillation phenomenon, and consequently, some degree of elasticity is needed in the sliding contact in order for stick-slip to occur. Israelachvili (1992) considered stick-slip on a molecular level, as measured with surface forces apparatus. He considers the order-disorder transformations described by Thompson and Robbins (1990, 1991) in terms of simulations. Most

Table 5.5 Reduction of static friction by surface films

Material combination	μ_s, No film	μ_s, Oxide film	μ_s, Sulfide film
Copper-on-copper	1.21	0.76	0.66
Steel on steel	0.78	0.27	0.39
Steel on steel	078	0.19[a]	0.16[a]

[a] Film and oil.

classical treatments of stick-slip take a mechanics approach, considering that the behaviour in unlubricated solid sliding is caused by forming and breaking adhesive bonds.

Stick-slip behaviour can be modelled in several ways. Generally, the system is represented schematically as a spring-loaded contact, sometimes including a dashpot element to account for viscoelastic response (Moore, 1975). The effects of time-dependent material properties on stick-slip behaviour of metals is provided by Kosterin and Kragelski (1962) and Kragelski (1965). Bowden and Tabor's analysis (1986) considers a free surface of inertial mass m being driven with a uniform speed v in the positive x direction against an elastic constant k. Then the instantaneous resisting force **F** over distance x equals $-kx$. With no damping of the resultant oscillation,

$$ma - -kx \quad [5.7]$$

Where acceleration $a=(d^2x/dt^2)$. The frequency n of simple harmonic motion is given by

$$n=(1/2\pi)(k/m)^{1/2} \quad [5.8]$$

Under the influence of a load **P** (mass **W** acting downward with the help of gravity g), the static friction force $\mathbf{F}_s$ can be represented as

$$\mathbf{F}_s = \mu_s \mathbf{P} \quad [5.9]$$

In terms of the deflection at the point of slip (x),

$$x = \mathbf{F}_s / k \quad [5.10]$$

If the kinetic friction coefficient μ is assumed to be constant during slip, then

$$ma - \mu\mathbf{P} = -kx \quad [5.11]$$

Letting time = 0 at the point of slip (where $x=\mathbf{F}_s/k$), and the forward velocity $v <<$ the velocity of slip, then,

$$x = (\mathbf{P}/k)\left[(\mu_s - \mu)\cos\omega t + \mu\right] \quad [5.12]$$

where, $\omega = (k/m)$.$^{1/2}$ In this case, the magnitude of slip, δ, is

$$\delta = \left[\mathbf{P}(2\mu_s - 2\mu)/k\right] \qquad [5.13]$$

From this equation, the larger the μ relative to μ_s, the less the effects of stick-slip, and when they are equal, the sliding becomes completely steady. Kudinov and Tolstoy (1986) derived a critical velocity above which stick-slip could be suppressed. This critical velocity v_c was directly proportional to the difference in the static and kinetic friction coefficients Δμ and inversely proportional to the square root of the product of the relative dissipation of energy during oscillation ($\psi = 4\pi\tau$), the stiffness of the system k, and the slider mass m. Thus,

$$v_c = \Delta\mu N/\sqrt{\psi k m} \qquad [5.14]$$

where N is the factor of safety. The authors report several characteristic values of $\Delta\mu$ for slideways on machine tools: cast iron on cast iron = 0.08, steel on cast iron = 0.05, bronze on cast iron = 0.02, and PTFE on cast iron = 0.04.

System resonance within limited stick-slip oscillation ranges was discussed by Bartenev and Lavrentev (1981), who cited experiments in which an oscillating normal load was applied to a system in which stick-slip was occurring. The minimum in stick-slip amplitude and friction force occurred over a range of about 1.5–2.5 kHz, the approximate value predicted by $(1/2\pi)(k/m)^{-1/2}$. Rabinowicz (1965) suggested two possible solutions:

1. Decrease the slip amplitude or slip velocity by increasing contact stiffness, increasing system damping, or increasing inertia.
2. Lubricate or otherwise form a surface film to ensure a positive μ versus velocity relationship.

The latter solution requires that effective lubrication be maintained, and stick-slip can return if the lubricant becomes depleted. The fact that stick-slip is associated with a significant difference between static and kinetic friction coefficients suggests that strategies that lower the former or raise the latter can be equally effective.

5.1.2 Sliding friction

Sliding friction plays a very important role in many manufacturing processes and sliding friction models, other than empirical models, can generally be grouped into five categories:

1. plowing and cutting-based models;
2. adhesion, junction-growth, and shear models;
3. single- and multiple-layer shear models;
4. debris layer and transfer layer models; and
5. molecular dynamics' models.

Each type of model was developed to explain frictional phenomena. Some of the models are based on observations that contact surfaces contain grooves that are suggestive of a dominant contribution from plowing, while single-layer models rely on a view of the interface showing flat surfaces separated by a layer whose shear strength controls friction. Some models involve combinations, such as adhesion plus plowing, while recent friction models contain molecular-level phenomena. Lubrication-oriented models and the debris-based models describe phenomena that take place in zone I, whereas most of the classical models for solid friction concern zone II phenomena. There are few models that take into account the effects of both the interfacial properties and the surrounding mechanical systems such as zone III models.

Models for sliding friction

Sliding friction models are summarized in this section of the chapter and fall into one, or more, of the five categories explained in the previous section.

(a) Plowing models

Plowing models assume that the dominant contribution to friction is the energy required to displace material ahead of a rigid protuberance or protuberances moving along a surface. One of the simplest models for plowing is that of a rigid cone of slant angle θ plowing through a surface under a normal load **P** (Rabinowicz, 1965). If we assign a groove width w (i.e., twice the radius r of the circular section of the penetrating cone at surface level), the triangular projected area, A_p, swept out as the cone moves along is as follows:

$$A_p = \frac{1}{2} w(r \tan\theta) = \frac{1}{2}(2r)(r \tan\theta) = r^2 \tan\theta \qquad [5.15]$$

The friction force $\mathbf{F}_p$ for this plowing contribution to sliding is found by multiplying the swept-out area by the compressive strength p. Thus, $\mathbf{F}_p = (r^2 \tan\theta)p$, and the friction coefficient, if this were the only contribution, is $\mu_p = \mathbf{F}_p/\mathbf{P}$. From the definition of the compressive strength p as force per unit area, we can write:

$$p = \mathbf{P}/\pi r^2 \qquad [5.16]$$

and

$$\mu_p = \mathbf{F}_p/\mathbf{P} = (r^2 \tan\theta)p/\pi r^2 p = (\tan\theta)/\pi \qquad [5.17]$$

This expression can also be written in terms of the apex angle of the cone α $(=90-\theta)$:

$$\mu_p = (2 \cot\alpha)/\pi \qquad [5.18]$$

Note that the friction coefficient calculated is for the plowing of a hard asperity and is not necessarily the same as the friction coefficient of the material sliding along the sides of the conical surface. Table 5.6 shows the maximum plowing contribution to friction for various metals.

(b) Adhesion, junction growth, and shear (AJS) models

The AJS interpretations of friction are based on a scenario in which two rough surfaces are brought close together, causing the highest peaks (asperities) to touch. As the normal force increases, the contact area increases and the peaks are flattened. Asperity junctions grow until they are able to support the applied load and adhesive bonds form at the contact points. When a tangential force is applied, the bonds must be broken, and overcoming the shear strength of the bonds results in the friction force. Early calculations comparing bond strengths to friction forces obtained in experiments raised questions as to the general validity of such models. Observations of material transfer and similar phenomena suggested that the adhesive bonds might be stronger than the softer of the two bonded materials, and that the shear strength of the softer material, not the bond strength, should be used in friction models.

Table 5.6 **Estimates of the maximum plowing contribution to friction**

Metal	Critical rake angle[a] (degrees)	μ_p
Aluminium	–5	0.03
Nickel	–5	0.03
Lead	–35	0.22
α-Brass	–35	0.22
Copper	–45	0.32

[a] For a cone, the absolute value of the critical rake angle is 90 minus angle θ.

Traditional friction models, largely developed for metal-on-metal sliding, have added the force contribution due to the shear of junctions to the contribution from plowing, giving the extended expression:

$$\mu = \left(\tau A_r\right) / \mathbf{P} + \left(\tan\theta\right) / \pi \quad [5.19]$$

where A_r is the real area of contact and τ is the shear strength of the material being plowed. This type of expression has met with relatively widespread acceptance in the academic community and is often used as the basis for other sliding friction models. But if the tip of the cone wears down, three contributions to the plowing process can be identified: the force needed to displace material from in front of the cone, the friction force along the leading face of the cone (i.e., the component in the macroscopic sliding direction), and the friction associated with shear of the interface along the worn frustum of the cone. From this analysis, it is clear that friction on two scales is involved: the macroscopic friction force for the entire system, and the friction forces associated with the flow of material along the face of the cones and across its frustum. That situation is somewhat analogous to the interpretation of orthogonal cutting of metals in which the friction force of the chip moving up along the rake face of the tool and friction along the wear land are not in general the same as the cutting force for the tool as a whole (Black, 1961). Considering the three contributions to the friction of a flat-tipped cone gives

$$\mu = \left(\tau / \mathbf{P}\right)\left(\pi r^2\right) + \mu_i \cos^2\theta + \left(\tan\theta\right) / \pi \quad [5.20]$$

where r is defined as the radius of the top of the worn cone and μ_i is the friction coefficient of the cone against the material flowing across its face. Equation [5.20] helps explain why the friction coefficients for ceramics and metals sliding on faceted diamond films are ten or more times higher than the friction coefficients reported for smooth surfaces of the same materials sliding against smooth surfaces of diamond (i.e., $\mu >> \mu_i$). When the rake angle θ is small, $\cos^2 \theta$ is close to 1.0, and the second term is only slightly less than μ_i (0.02–0.12 typically). If one assumes that the friction coefficient for the material sliding across the frustum of the cone is the same as that for sliding along its face (μ_i), then eq. [5.20] can be re-written:

$$\mu = 2\mu_i + (\tan \theta) / \pi \quad [5.21]$$

Thus, implying that the friction coefficient for a rigid sliding cone is more than twice that for sliding a flat surface of the same two materials. It is interesting to note that eq. [5.21] does not account for the depth of penetration, a factor that seems critical for accounting for the energy required to plow through the surface (displace the volume of material ahead of the slider), and at $\theta = 90°$, which implies infinitely deep penetration of the cone, it would be impossible to move the slider at all as μ tends towards infinity.

When one views the complexities of surface finish it seems remarkable that eqs. [5.20] and [5.21], which depend on a single quantity [$(\tan \theta)/\pi$], should be able to predict the friction coefficient with any degree of accuracy. The model is based on a single conical asperity cutting through a surface that makes no obvious accountability for multiple contacts and differences in contact angle. The model is also based on a surface's relatively ductile response to a perfectly rigid asperity and can neither account for fracture during wear nor account for the change in the groove geometry that one would expect for multiple passes over the same surface.

Mulhearn and Samuels (1962) published a paper on the transition between abrasive asperities cutting through a surface and plowing through it. The results of their experiments suggested that there exists a critical rake angle for that type of transition. (Note: The rake angle is the angle between the normal to the surface and the leading face of the asperity, with negative values indicating a tilt toward the direction of travel.) If plowing can occur only up to the critical rake angle, then we may compute the maximum contribution to friction due to plowing from the data of Mulhearn and Samuels and eq. [5.18] (Table 5.6). This approach suggests that the maximum contribution of plowing to the friction coefficient of aluminium or nickel is about 0.03 in contrast to copper, whose maximum plowing contribution is 0.32. Since the sliding friction coefficient for aluminium can be quite high (over 1.0 in some cases), the implication is that factors other than plowing, such as the shearing of strongly adhering junctions, would be the major contributor. Examination of unlubricated sliding wear surfaces of both Al and Cu often reveals a host of ductile-appearing features not in any way resembling cones, and despite the similar appearances in the microscope of worn Cu and Al, one finds from the first and last rows in Table 5.6 that the contribution of plowing to friction should be different by a factor of 10. Again, the simple cone model appears to be too simple to account for the difference.

Hokkirigawa and Kato (1988) carried the analysis of abrasive contributions to sliding friction even further using observations of single hemispherical sliding contacts (quenched steel, tip radius 26 or 62 μm) on brass, carbon steel, and stainless steel in a scanning electron microscope. They identified three modes: (a) plowing, (b) wedge formation and (c) cutting (chip formation). The tendency of the slider to produce the various modes was related to the degree of penetration, D_p. Here, $D_p = h/a$, where h is the groove depth and a is the radius of the sliding contact. The sliding friction coefficient was modelled in three ways depending upon the regime of sliding. Three parameters were introduced: $f = p/\tau$,

$\theta = \sin^{-1}(a/R)$ and β, the angle of the stress discontinuity (shear zone) from Challen and Oxley's (1979) analysis. Where p is the contact pressure, τ is the bulk shear stress of the flat specimen, and R is the slider tip radius. The friction coefficient was given as follows for each mode:

Cutting mode:

$$\mu = \tan\left[\theta - (\pi/4) + \frac{1}{2}\cos^{-1} f\right] \qquad [5.22]$$

Wedge-forming mode:

$$\mu = \frac{\left\{1 - \sin 2\beta + \left(1 - f^2\right)^{1/2}\right\} \sin\theta + f\cos\theta}{\left\{1 - \sin 2\beta + \left(1 - f^2\right)^{1/2}\right\} \cos\theta + f\sin\theta} \qquad [5.23]$$

Plowing mode:

$$\mu = \frac{A\sin\theta + \cos\left(\cos^{-1} f - \theta\right)}{A\sin\theta + \cos\left(\cos^{-1} f - \theta\right)} \qquad [5.24]$$

where

$$A = 1 + (\pi/2) + \cos^{-1} f - 2\theta - 2\sin^{-1}\frac{\sin\theta}{(1-f)^{-1/2}} \qquad [5.25]$$

For unlubricated conditions, the transitions between the various modes were experimentally determined by observation in the scanning electron microscope. Table 5.7 summarizes those results. Results of the study illustrate the point that the analytical form of the frictional dependence on the shape of asperities cannot ignore the mode of surface deformation. In summary, the foregoing treatments of the plowing contribution to friction assumed that asperities could be modelled as regular geometric shapes. However, rarely do such shapes appear on actual sliding surfaces. The

Table 5.7 **Critical degree of penetration (D_p) for unlubricated friction mode transitions**

	Value of D_p for the transition	
Material	Plowing to wedge formation	Wedge formation to cutting
Brass	0.17 (tip radius 62 μm)	0.23 (tip radius 62, 27 μm)
Carbon steel	0.12 (tip radius 62 μm)	0.23 (tip radius 27 μm)
Stainless steel	0.13 (tip radius 62, 27 μm)	0.26 (tip radius 27 μm)

asperities present on most sliding surfaces are irregular in shape, as viewed with a scanning electron microscope.

(c) Plowing with debris generation

Even when the predominant contribution to friction is initially from cutting and plowing of hard asperities through the surface, the generation of wear debris that submerges the asperities can reduce the severity of plowing. Table 5.8 shows that starting with multiple hard asperities of the same geometric characteristics produced different initial and steady-state friction coefficients for the three slider materials. Wear debris accumulation in the contact region affected the

Table 5.8 **Effects of material type on friction during abrasive sliding[a]**

	24 μm grit size		16 μm grit size	
Slider material	Starting μ	Ending μ	Starting μ	Ending μ
AISI 52100 steel	0.47	0.35	0.45	0.29
2014-T4 aluminium	0.69	0.56	0.64	0.62
PMMA	0.73	0.64	0.72	0.60

[a] Normal force 2.49 N, sliding speed 5 mm/sec, multiple strokes 20 mm long.

frictional behaviour. In the case of abrasive papers and grinding wheels, this is called loading and is extremely important in grinding, and a great deal of effort has been focused on dressing grinding wheels to improve their material removal efficiency. One measure of the need for grinding wheel dressing is an increase in the tangential grinding force or an increase in the power drawn by the grinding spindle.

As wear progresses, the wear debris accumulates between the asperities and alters the effectiveness of the cutting and plowing action by covering the active points. If the cone model is to be useful at all for other than pristine surfaces, the effective value of θ must be given as a function of time or number of sliding passes. Not only is the wear rate affected, but the presence of debris affects the interfacial shear strength, as is explained later in this chapter in regard to third-body particle effects on friction. The observation that wear debris can accumulate and so affect friction has led investigators to try patterning surfaces to create pockets where debris can be collected (Suh, 1986). The orientation and depths of the ridges and grooves in a surface affect the effectiveness of the debris-trapping mechanism.

(d) Plowing with adhesion

Traditional models for sliding friction have historically been developed with metallic materials in mind. Classically, the friction force is said to be an additive contribution of adhesive (S) and plowing forces ($\mathbf{F}_{pl}$) (Bowden and Tabor, 1986):

$$\mathbf{F} = \mathbf{S} + \mathbf{F}_{pl} \qquad [5.26]$$

The adhesive force derives from the shear strength of adhesive metallic junctions that are created when surfaces touch one another under a normal force. Thus, by dividing by the normal force we find that $\mu = \mu_{\text{adhesion}} + \mu_{\text{plowing}}$. If the shear strength of the junction is τ and the contact area is A, then

$$\mathbf{S} = \tau A \tag{5.27}$$

The plowing force $\mathbf{F}_{pl}$ is given by

$$\mathbf{F}_{pl} = pA' \tag{5.28}$$

where p is the mean pressure to displace the metal in the surface and A' is the cross section of the grooved wear track. While helpful in understanding the results of experiments in the sliding friction of metals, the approach involves several applicability-limiting assumptions: for example, that adhesion between the surfaces results in bonds that are continually forming and breaking; that the protuberances of the harder of the two contacting surfaces remain perfectly rigid as they plow through the softer counterface; and perhaps most limiting of all, that the friction coefficient for a tribosystem is determined only from the shear strength properties of materials.

(e) Single-layer shear (SLS) models

The SLS models for friction depict an interface as a layer whose shear strength determines the friction force, and hence, the friction coefficient. The layer can be a separate film, like a solid lubricant, or simply the near surface zone of the softer material that is shearing during friction. The friction force **F** is the product of the contact area A and the shear strength of the layer:

$$\mathbf{F} = \tau A \tag{5.29}$$

The concept that the friction force is linearly related to the shear strength of the interfacial material has a number of useful implications, especially as regards the role of thin lubricating layers, including oxides and tarnish films. It is known from the work of Bridgman (1931) on the effects of pressure on mechanical properties that τ is affected by contact pressure, p:

$$\tau = \tau_o + \alpha p \qquad [5.30]$$

Table 5.9 lists several values for the shear stress and the constant α (Kragelskii et al., 1982).

(f) Multiple-layer shear (MLS) models

The MLS models presume that the sliding friction can be explained on the basis of the shear strength on a single layer interposed between solid surfaces. Evidence revealed by the examination of frictional surfaces suggests that shear can occur at various positions in the interface: for example, at the upper interface between the solid and the debris layer, within the entrapped debris or transfer layer itself, at the lower interface, or even below the original surfaces where extended delaminations may occur. Therefore, one may construct a picture of sliding friction that involves a series of shear layers (sliding resistances) in parallel. Certainly, one would expect the predominant frictional contribution to be the lowest shear strength in the shear layers. Yet the shear

Table 5.9 Measured values for the shear stress dependence on pressure

Material	τ_o (kgf/mm^2)	μ
Aluminium	3.00	0.043
Beryllium	0.45	0.250
Chromium	5.00	0.240
Copper	1.00	0.110
Lead	0.90	0.014
Platinum	9.50	0.100
Silver	6.50	0.090
Tin	1.25	0.012
Vanadium	1.80	0.250
Zinc	8.00	0.020

forces transmitted across the weakest interface may still be sufficient to permit some displacement to occur at one or more of the other layers above or below it, particularly if the difference in shear strengths between those layers is small.

The MLS models can be treated like electrical resistances in a series. The overall resistance of such a circuit is less than any of the individual resistances because multiple current paths exist. Consider, for example, the case where there are three possible operable shear planes stacked up parallel to the sliding direction in the interface. Then

$$\frac{1}{\mathbf{F}} = \frac{1}{\mathbf{F}_1} + \frac{1}{\mathbf{F}_2} + \frac{1}{\mathbf{F}_3} \qquad [5.31]$$

And, solving for the total friction force **F**, in terms of the friction forces acting on the three layers, is

$$\mathbf{F} = \frac{\mathbf{F}_1\,\mathbf{F}_2\,\mathbf{F}_3}{\mathbf{F}_1\,\mathbf{F}_2 + \mathbf{F}_2\,\mathbf{F}_3 + \mathbf{F}_1\,\mathbf{F}_3} \qquad [5.32]$$

If the area of contact A is the same across each layer, then eq. [5.32] can be written in terms of the friction coefficient of the interface, the shear stresses of each layer, and the normal load **P** as follows:

$$\mu = \left(\frac{A}{\mathbf{P}}\right)\left[\frac{\tau_1\,\tau_2\,\tau_3}{\tau_1\,\tau_2 + \tau_2\,\tau_3 + \tau_1\,\tau_3}\right] \qquad [5.33]$$

If one of the shear planes suddenly became unable to deform (say, by work hardening or by clogging with a compressed clump of wear debris), the location of the governing plane of shear may shift quickly, causing the friction to fluctuate. Thus, by writing the shear stresses of each layer as functions of time, the MLS model has the advantage of being able to account for variations in friction force with time and may account for some of the features observed in microscopic examinations of wear tracks.

(g) Molecular dynamics' models

When coupled with information from nanoprobe instruments, such as the atomic force microscope, the scanning tunnelling microscope, the surface-forces apparatus, and the lateral-force microscope, MD studies have made possible insights into the behaviour of pristine surfaces on the atomic scale. Molecular dynamics models of friction for assemblages of even a few hundred atoms tend to require millions upon millions of individual, iterative computations to predict frictional interactions taking place over only a fraction of a second in real time. Because they begin with very specific arrangements of atoms, usually in single crystal form with a specific sliding orientation, results are often periodic with sliding distance. Some of the calculation results are remarkably similar to certain types of behaviour observed in real materials, simulating such phenomena as dislocations (localized slip on preferred planes) and the adhesive transfer of material to the opposing counterface. However, molecular dynamics models are not at present capable of handling such contact surface features as surface fatigue-induced delaminations, wear debris particles compacting and deforming in the interface, high-strain-rate phenomena, work hardening of near-surface layers, or effects of inclusions and other artefacts present in the microstructures of commercial engineering materials.

The models presented up to this point use either interfacial geometric parameters or materials properties (i.e., bonding energies, shear strengths, or other mechanical properties) to predict friction. Clearly, frictional heating and the chemical environment may affect some of the variables used in these models. For example, the shear strength of many metals decreases as the temperature increases and increases as the speed of deformation increases. Certainly, wear and its consequences (debris) will affect friction. Thus, any of the previously described models will probably require some sort of modification, depending on the actual conditions of sliding contact. In general, the following can be said about friction models:

- No existing friction model explicitly accounts for all the possible factors that can affect friction.
- Even very simple friction models may work to some degree under well-defined, limited ranges of conditions, but their applicability must be tested in specific cases.
- Accurately predictive, comprehensive tribosystem-level models that account for interface geometry, materials properties, lubrication aspects, thermal, chemical, and external mechanical system response, all in a time-dependent context, do not exist.
- Friction models should be selected and used based on an understanding of their limitations and on as complete as possible an understanding of the dominant influences in the tribosystem to which the models will be applied.
- Current quantitative models produce a single value for the friction force, or friction coefficient. Since the friction force in nearly all known tribosystems varies to some degree, any model that predicts a single value is questionable.

If no existing model is deemed appropriate, the investigator could either modify a current model to account for the additional variables, develop a new system-specific model, or revert to simulative testing and/or field experiments to obtain the approximate value. An alternative to modelling is to estimate frictional behaviour using a graphical, or statistical approach.

5.1.3 Frictional heating

Heat generation and rising surface temperatures are intuitively associated with friction. When a friction force **F** moves through a distance x, an amount of energy $\mathbf{F}x$ is produced. The laws of thermodynamics require that the energy so produced be dissipated to the surroundings. At equilibrium, the energy into a system U_{in} equals the sum of the energy output to the surroundings U_{out} (dissipated externally) and the energy accumulated $U_{accumulated}$ (consumed or stored internally):

$$U_{in} = U_{out} + U_{accumulated} \qquad [5.34]$$

The rate of energy input in friction is the product of **F** and the sliding velocity v whose units work out to energy per unit time (e.g., Nm/sec). This energy input rate at the frictional interface is balanced almost completely by heat conduction away from the interface, either into the contacting solids or by radiation or convection to the surroundings. In general, only a small amount of frictional energy, perhaps only 5%, is consumed or stored in the material as microstructural defects such as dislocations, the energy to produce phase transformations, surface energy of new wear particles and propagating subsurface cracks, etc. Most of the frictional energy is dissipated as heat, and under certain conditions there is enough heat to melt the sliding interface. Energy that cannot readily be conducted away from the interface raises the temperature locally. Assuming that the proportionality of friction force **F** to normal force **P** (i.e. by definition, $\mathbf{F} = \mu\mathbf{P}$) holds over a range of normal forces, we would expect that the temperature rise in a constant-velocity sliding system should increase linearly with the normal force. Tribologists distinguish between two temperatures, the flash temperature and the mean surface temperature. The former is localized, the latter averaged out over the nominal contact zone. Since sliding surfaces touch at only a few locations at any instant, the energy is concentrated there and the heating is particularly intense – thus the name flash temperature. The combined effect of many such flashes dissipating their energy in the interface under steady state is to heat a near-surface layer to an average temperature that is determined by the energy transport conditions embodied in eq. [5.34] given earlier. Blok (1963) discussed the concept and calculation of flash temperature in a review article. The early work of Blok (1937) and Jaeger (1942) is still cited as a basis for more recent work, and it has been reviewed in a simplified form by Bowden and Tabor (1986). Basically, the temperature rise in the interface is given as a function of the total heat developed, Q:

$$Q = \frac{\mu \mathbf{W} g v}{J} \qquad [5.35]$$

where μ is the sliding friction coefficient, $\mathbf{W}$ is the load, g the acceleration due to gravity, v the sliding velocity, and J the mechanical equivalent of heat (4.186 J/cal). Expressions for various heat flow conditions are then developed based on eq. [5.35]. Some of these are given in Table 5.10, which shows the expressions become more complicated when the cooling effects of the incoming, cooler surface are accounted for. Rabinowicz (1965) published an expression for estimating the flash temperature rise in sliding:

$$\theta_m = \frac{v}{2}(\pm \text{ a factor of 2 to 3}) \qquad [5.36]$$

where v is sliding velocity (ft/min) and θ_m is the estimated surface flash temperature (°F). A comparison of the results of using eq. [5.36] with several other, more complicated models for frictional heating has provided similar results, but more rigorous treatments are sometimes required to account for

Table 5.10 Temperature rise during sliding

Conditions	Temperature rise[a] ($T = T_o$)
Circular junction of radius a	$= \dfrac{Q}{4a(k_1 + k_2)}$
Square junction of side = 2*l*, at low speed	$= \dfrac{Q}{4.24l(k_1 + k_2)}$
Square junction of side = 2*l*, at high speed wherein the slider is being cooled by the incoming surface of the flat disk specimen	$= \dfrac{Qx^{1/2}}{3.76l\left[k_1 (l/v)^{1/2} + 1.125x^{1/2}k_2\right]}$ Where $x = (k_1/\rho_1 c_1)$ for the disk material

[a] Key: T = steady-state junction temperature, T_o = initial temperature, $k_{1,2}$ = thermal conductivity of the slider and flat bodies, ρ = density, *c* = specific heat.
After Jaeger (1942).

the variables left out of this rule of thumb. In general, nearly all models for flash or mean temperature rise during sliding contain the friction force-velocity product. Sometimes, the friction force is written as the product of the normal force and friction coefficients.

A review of frictional heating calculations has been provided by Cowan and Winer (1992), along with representative materials properties data to be used in those calculations. Their approach involves the use of two heat partition coefficients (γ_1 and γ_2) that describe the relative fractions of the total heat that go into each of the contacting bodies, such that $\gamma_1 + \gamma_2 = 1$. The time that a surface is exposed to frictional heating will obviously affect the amount of heat it receives. The Fourier modulus, F_o, a dimensionless parameter, is introduced to establish whether or not steady-state conditions have been reached at each surface. For a contact radius *a*, an exposure time *t*, and a thermal diffusivity for body *i* of D_i,

$$F_o = \frac{D_i t}{a^2} \quad [5.37]$$

The Fourier modulus is taken to be 100 for a surface at steady state conditions. Another useful parameter grouping is the Peclet number P_e, defined in terms of the density of the solid ρ, the specific heat c_p, the sliding velocity v, the thermal conductivity k, and the characteristic length L_c:

$$P_e = \frac{\rho c_p v L_c}{k} \quad [5.38]$$

The characteristic length is the contact width for a line contact or the contact radius for a circular contact. The Peclet number relates the thermal energy removed by the surrounding medium to that conducted away from the region in which frictional energy is being dissipated. As $D_i = (\rho c_p / k)$ yields the following,

$$P_e = \frac{v L_c}{D_i} \quad [5.39]$$

the Peclet number is sometimes used as a criterion for determining when to apply various forms of frictional heating models. It is also used in understanding frictional heating problems associated with grinding and machining processes.

It is important to compare the forms of models derived by different authors for calculating flash temperature rise. Four treatments for a pin moving along a stationary flat specimen are briefly compared: Rabinowicz's derivation based on surface energy considerations, a single case from Cowan and Winer's review, Kuhlmann-Wilsdorf's model, and the model provided by Ashby. Based on considerations of junctions of radius r and surface energy of the softer material Γ, Rabinowicz arrived at the following expression:

$$T_f = \frac{3000\ \pi\mu\Gamma v}{J(k_1 + k_2)} \qquad [5.40]$$

where J is the mechanical equivalent of heat, v is sliding velocity, μ is the friction coefficient, and k_1 and k_2 represent the thermal conductivities of the two bodies. The constant 3000 obtained from the calculation of the effective contact radius r in terms of the surface energy of the circular junctions Γ and their hardness H (i.e., $r = 12{,}000\Gamma H$) and the load carried by each asperity ($\mathbf{P} = \pi r^2 H$). Thus, the numerator is actually the equivalent of $\mathbf{F}v$ expressed in terms of the surface energy model. The equation provided by Cowan and Winer, for the case of a circular contact with one body in motion, is

$$T_f = \frac{\gamma_1 \mu \mathbf{P} v}{\pi a k_1} \qquad [5.41]$$

where γ_1 is the heat partition coefficient, described earlier, **P** is the normal force, a is the radius of contact, and k_1 is as defined earlier. The value of γ_1 takes various forms depending on the specific case. The presence of elastic, or plastic, contact can also affect the form of the average flash temperature, as Table 5.11 demonstrates. Here, the exponents of normal force and velocity are not unity in all cases.

Kuhlmann-Wilsdorf (1987) considered an elliptical contact area as the planar moving heat source. The flash temperature is given in terms of the average temperature in the interface T_{ave}:

$$T_{ave} = \frac{\pi q r}{4k_1} \qquad [5.42]$$

where q is the rate of heat input per unit area (related to the product of friction force and velocity), r is the contact spot radius, and k_1 is the thermal conductivity, as given earlier.

Table 5.11 **Effects of deformation type and Peclet number on flash temperature calculation for the circular contact case**

Type of deformation	Peclet number	Average flash temperature[a]
Plastic	$P_e < .02$	$T_f = \mu P^{0.5} v \frac{\sqrt{\pi\rho}}{8k}$
Plastic	$P_e > 200$	$T_f = 0.31\mu \mathbf{P}^{0.25} v^{0.5} \left[\frac{(\pi\rho)^{0.75}}{(k\rho C)^{0.5}}\right]$
Elastic	$P_e < .02$	$T_f = 0.13\mu \mathbf{P}^{0.667} v \left(\frac{1}{k}\right)\left(\frac{E_v}{R}\right)^{0.333}$
Elastic	$P_e > 200$	$T_f = 0.36\mu \mathbf{P}^{0.5} v^{0.5} \left(\frac{1}{\sqrt{k\rho c}}\right)\left(\frac{E_v}{R}\right)^{0.5}$

[a] Key: μ = friction coefficient, **P** = load, v = velocity, k = thermal conductivity, π = density, c = heat capacity, E_v = the reduced elastic modulus = $E/(1-v^2)$, v = Poisson's ratio, ρ = flow pressure of the softer material.

Then

$$T_f = \frac{T_{avc}}{(1/ZS) + (k_1/S_0)} \quad [5.43]$$

where Z is a velocity function and S and S_o are contact area shape functions (both = 1.0 for circular contact). At low speeds, where the relative velocity of the surfaces $v_r < 2$ ($v_r = v/\mathrm{P}_e$), Z can be approximated by $1/[1 + (v_r/3)]$. The differences between models for frictional heating arise from the following:

- assuming different shapes for the heat source on the surface;
- different ways to partition the flow (dissipation) of heat between sliding bodies;

- different ways to account for thermal properties of materials (e.g. using thermal diffusivity instead of thermal conductivity, etc.);
- different contact geometry (sphere-on-plane, flat-on-flat, cylinder-on-flat, etc.);
- assuming heat is produced from a layer (volume) instead of a planar area; and
- changes in the form of the expression as the sliding velocity increases.

Comparing the temperature rises predicted by different models for low sliding speeds produces accurate results, even with the uncertainties in the values of the material properties that go into the calculations. At higher speeds, the predictions become unreliable since materials properties change as a function of temperature and the likelihood of the interface reaching a steady state is much lower. Experimental studies have provided very useful information in validating the forms of frictional heating models. Experimental scientists have often used embedded thermocouples in one or both members of the sliding contact to measure surface temperatures, and others sometimes made thermocouples out of the contacts themselves. However, techniques using infrared sensors have been used as well. Dow and Stockwell (1977) used infrared detectors with a thin, transparent sapphire blade sliding on a 15-cm-diameter ground cylindrical drum to study the movements and temperatures of hot spots. Griffioen et al. (1985) and Quinn and Winer (1987) used an infrared technique with a sphere-on-transparent sapphire disk geometry. A similar arrangement was also developed and used by Furey with copper, iron, and silver spheres sliding on sapphire, and Enthoven et al. (1993) used an infrared system with a ball-on-flat arrangement to study the relationship between scuffing and the critical temperature for its onset.

Frictional heating is important because it changes the shear strengths of the materials in the sliding contact, promotes reactions of the sliding surfaces with chemical species in the environment, enhances diffusion of species, and can result in the breakdown or failure of the lubricant to perform its functions. Under extreme conditions, such as plastic extrusion, frictional heating can result in molten layer formation that serves as a liquid lubricant.

5.2 Lubrication to control friction in manufacturing

The frictional characteristics of liquid and solid lubricants and their interaction with materials are reviewed, while comprehensive discussions of the mechanical and chemical engineering aspects of lubrication are available in the literature (Wills, 1980).

5.2.1 Liquid lubrication

The process of lubrication is one of supporting the contact pressure between opposing surfaces, helping to separate them, and at the same time reducing the sliding or rolling resistance in the interface. There are several ways to accomplish this. One way is to create in the gap between the bodies geometric conditions that produce a fluid pressure sufficient to prevent the opposing asperities from touching while still permitting shear to be fully accommodated within the fluid. That method relies on fluid mechanics and modifications of the lubricant chemistry to tailor the liquid's properties. Another way to create favourable lubrication conditions is to formulate the liquid lubricant in such a way that chemical species within it react with the surface of the bodies to form shearable solid films. Surface species need not react with the lubricant, but catalyse the reactions that produce these protective films.

Several attributes of liquids make them either suitable or unsuitable as lubricants. Klaus and Tewksbury (1984) have discussed these characteristics in some detail. They include:

- density;
- bulk modulus;
- gas solubility;
- foaming and air entrainment tendencies;
- viscosity and its relationships to temperature and pressure;
- vapour pressure;
- thermal properties and stability; and
- oxidation stability.

The viscosity of fluids usually decreases with temperature and therefore can reduce the usefulness of a lubricant as temperature rises. The term viscosity index, abbreviated VI, is a means to express this variation. The higher the VI, the less the change in viscosity with temperature. One of the types of additives used to reduce the sensitivity of lubricant viscosity to temperature changes is called a VI improver. ASTM test method D 2270 is one procedure used to calculate the VI and the process is described step-by-step in the article by Klaus and Tewksbury (1984). The method involves references to two test oils, the use of two different methods of calculation (depending on the magnitude of VI), and relies on charts and tables.

ASTM Standard D341 recommends using the Walther equation to represent the dependence of lubricant viscosity on temperature. Defining Z as the viscosity in cSt plus a constant T (typically ranging from 0.6 to 0.8 with ASTM specifying 0.7) equal to the temperature in Kelvin or Rankin, and A and B being constants for a given oil, then

$$\log_{10}(\log_{10} Z) = A + B(\log_{10} T) \quad [5.44]$$

Sanchez-Rubio et al. (1992) have suggested an alternative method in which the Walther equation is used. In this case, they define a viscosity number (VN) as follows:

$$\mathrm{VN} = \left[1 + \frac{(3.55 + \mathrm{B})}{3.55}\right] \times 100 \quad [5.45]$$

The value of 3.55 was selected because lubricating oils with a VI of 100 have a value of B about equal to −3.55. Using this expression implies that VN = 200 would correspond to an idealized oil whose viscosity has no dependence of viscosity on temperature (i.e. $B = 0$). The pressure to which an oil is subjected to can influence its viscosity, so the relationship between dynamic viscosity and hydrostatic pressure p can be represented by

$$\eta = \eta_0 \exp(\alpha p) \quad [5.46]$$

where η and α vary with the type of oil. Table 5.12 illustrates the wide range of viscosities possible for several liquid lubricants under various temperatures and pressures. The viscosity indices for these oils range from −132 to 195. Viscosity has a large effect on determining the regime of lubrication and the resultant friction coefficient. Similarly to the effect of strain rate on the shear strength of certain metals, like aluminium, the rate of shear in the fluid can also alter the viscosity of a lubricant.

Ramesh and Clifton (1987) constructed a plate impact device to study the shear strength of lubricants at strain rates as high as 900,000/sec and found significant effects of shear

Table 5.12 **Effects of temperature and pressure on viscosity of selected lubricants having various viscosity indexes**

Quantity	Fluorolube	Hydrocarbon	Ester	Silicone
Viscosity index	−132	100	151	195
Viscosity (cSt) at −40°C	500,000	14,000	3600	150
Viscosity (cSt) at −100°C	2.9	3.9	4.4	9.5
Viscosity (cSt) at −40°C and 138 MPa	2700	340	110	160
Viscosity (cSt) at −40°C and 552 MPa	>1,000,000	270,000	4900	48,000

Note: All fluids have viscosities of 20 cSt at 40°C and 0.1 MPa pressure.

rate on the critical shear stress of lubricants. In a Newtonian fluid, the ratio of shear stress to shear strain does not vary with stress, but there are other cases, such as for greases and solid dispersions in liquids, where the viscosity varies with the rate of shear. Such fluids are termed non-Newtonian and the standard methods for measuring viscosity cannot be used.

Lubrication regimes determine the effectiveness of fluid film formation, and hence, surface separation. In the first decade of the twentieth century, Stribeck developed a systematic method to understand and depict regimes of journal bearing lubrication, linking the properties of lubricant viscosity (η), rotational velocity of a journal (ω), and contact pressure (p) with the coefficient of friction. Based on the work of Mersey, McKee, and others, the dimension-less group of parameters has evolved into the more recent notation (ZN/p), where Z is viscosity, N is rotational speed, and p is pressure. The Stribeck curve has been widely used in the design of bearings and to explain various types of behaviour in the field of lubrication. At high pressures, or when the lubricant viscosity and/or speed are very low, surfaces may touch, leading to high friction. In that case, friction coefficients are typically in the range of 0.5–2.0. The level plateau at the left of the curve represents the boundary lubrication regime in which friction is lower than for unlubricated sliding contact ($\mu = 0.05$ to about 0.15). The drop-off in friction is called the mixed film regime. The mixed regime refers to a combination of boundary lubrication with hydrodynamic or elastohydrodynamic lubrication. Beyond the minimum in the curve, hydrodynamic and elastohydrodynamic lubrication regimes are said to occur. Friction coefficients under such conditions can be very low. Typical friction coefficients for various types of rolling element bearings range between 0.001 and 0.0018.

The conditions under which a journal bearing of length L, diameter D, and radial clearance C (bore radius minus bearing shaft radius) operates in the hydrodynamic regime can be summarized using a dimensionless parameter known as the Sommerfeld number S, defined by

$$S = \frac{\eta NLD}{P}\left(\frac{R}{C}\right)^2 \qquad [5.47]$$

where P is the load on the bearing perpendicular to the axis of rotation, N is the rotational speed, η is the dynamic viscosity of the lubricant, and R is the radius of the bore. The more concentrically the bearing operates, the higher the value of S, but as S approaches 0, the lubrication may fail, leading to high friction. Sometimes Stribeck curves are plotted using S instead of (ZN/p) as the abscissa. Raimondi and his co-workers (1968) added leakage considerations when they developed design charts in which the logarithm of the Sommerfeld number is plotted against the logarithm of either the friction coefficient or the dimensionless film thickness.

Using small journal bearings, McKee developed the following expression for the coefficient of friction μ based on the journal diameter D, the diametral clearance C, and an experimental variable k, which varies with the length to diameter ratio (L/D) of the bearing (Hall et al. 1961):

$$\mu = \left(4.73 \times 10^{-8}\right)\left(\frac{ZN}{p}\right)\left(\frac{D}{C}\right) + k \qquad [5.48]$$

The value of k is about 0.015 at $(L/D) = 0.2$, drops rapidly to a minimum of about 0.0013 at $(L/D) = 1.0$, and rises nearly linearly to about 0.0035 at $(L/D) = 3.0$. A simpler expression, discussed by Hutchings (1992), can be used for bearings that have no significant eccentricity:

$$\mu = \frac{2\pi}{S}\frac{h}{R} \qquad [5.49]$$

where S is the Sommerfeld number, h is the mean film thickness, and R is the journal radius. With good hydrodynamic lubrication and good bearing design, μ can be as low as 0.001.

Hydrodynamic lubrication, sometimes called thick-film lubrication, generally depends on the development of a converging wedge of lubricant in the inlet of the interface. This wedge generates a pressure profile to force the surfaces apart. When the elastic deformation of the solid bodies is similar in extent to the thickness of the lubricant film, then elastohydrodynamic lubrication is said to occur. This latter regime is common in rolling element bearings and gears where high Hertz contact stresses occur. If the contact pressure exceeds the elastic limit of the surfaces, plastic deformation and increasing friction occur. One way to understand and control the various lubrication regimes is by using the specific film thickness (also called the lambda ratio), defined as the ratio of the minimum film thickness in the interface (h) to the composite root-mean-square (rms) surface roughness σ^*:

$$\wedge = h / \sigma^* \tag{5.50}$$

where the composite surface roughness is defined in terms of the rms roughness ($\sigma_{1,2}$) of surfaces 1 and 2, respectively:

$$\sigma^* = \sqrt{\left(\sigma_1^2 + \sigma_2^2\right)} \tag{5.51}$$

For the boundary regime, $\Lambda << 1$; for the mixed regime, $1 < \Lambda < 3$; for the hydrodynamic regime, $\Lambda >> 6$; and for the elastohydrodynamic regime, $3 < \Lambda < 10$. Boundary lubrication produces friction coefficients that are lower than those for unlubricated sliding but higher than those for effective hydrodynamic lubrication, typically in the range $0.05 < \mu < 0.2$. Briscoe and Stolarski (1993) have reviewed friction under boundary-lubricated conditions. They cited the earlier work of Bowden, which gave the following expression for the friction coefficient under conditions of boundary lubrication:

$$\mu = \beta \mu_{\mathrm{a}} + (1 - \beta)\mu_1 \tag{5.52}$$

where the adhesive component μ_{a} and the viscous component of friction μ_1 are given in terms of the shear stress of the adhesive junctions in the solid (metal) τ_m and the shear strength of the boundary film τ_1 under the influence of a contact pressure σ_p:

$$\mu_a = \frac{\tau_m}{\sigma_p} \tag{5.53}$$

The parameter, β, is called the fractional film defect (Briscoe and Stolarski, 1993) and is:

$$\beta = 1 - \exp\left\{ -\left[\frac{\left(30.9 \times 10^5\right) \; T_{\mathrm{m}}^{1/2}}{VM^{1/2}} \right] \exp\left(-\frac{E_c}{RT} \right) \right\} \tag{5.54}$$

where M is the molecular weight of the lubricant, V is the sliding velocity, T_m is the melting temperature of the lubricant, E_c is the energy to desorb the lubricant molecules, R is the universal gas constant, and T is the absolute temperature. Various graphical methods have been developed to help select boundary lubricants and to help simplify the task of bearing designers. Most of these methods are based on the design parameters of bearing stress (or normal load) and velocity. One method, developed by Glaeser and Dufrane (1978) involves the use of design charts for different bearing materials. An alternate but similar approach was used in developing the so-called IRG transitions diagrams (subsequently abbreviated ITDs), an approach that evolved in the early 1980s, was applied to various bearing steels, and is still being used to define the conditions under which boundary-lubricated tribosystems operate effectively. Instead of pressure, load is plotted on the ordinale. Three regions of ITDs are defined in terms of their frictional behaviour: Region I, in which the friction trace is relatively low and smooth; Region II, in which the friction trace begins with a high level then settles down to a lower, smoother level; and Region III, in which the friction trace is irregular and remains high. The transitions between Regions I and II or between Regions I and III are described as a collapse of liquid film lubrication. The locations of these transition boundaries for steels were seen to depend more on the surface roughness of the materials and the composition of the lubricants and less on microstructure and composition of the alloys. Any of the following testing geometries can be used to develop ITDs: four-ball machines, ball-on-cylinder machines, crossed-cylinders machines, and flat-on-flat testing machines (including flat-ended pin-on-disk). One important aspect of the use of liquid lubricants is how they are applied, filtered, circulated, and replenished. Lubricants can also be formed on surfaces by the chemical reaction of vapour-phase precursor species in argon and nitrogen environments.

5.2.2 Liquid lubricant compositions

Most lubricating oils in use are petroleum based and are obtained by refining distillate from residual fractions obtained from crude oil. Lubricating base oils are complex mixtures of multiple-ring molecules with side chains attached. Lubricating oil contains aromatic rings, naphthenic rings, and side chains. Consequently, base oils are classified as paraffinic, naphthenic, or aromatic, depending on their molecular structures, the length of the side chains, and the ratio of carbon atoms in the side chains to those in the rings.

Zisman (1959) conducted experiments on monomolecular films on glass to illustrate the effect of carbon chain length on friction. Above 14 atoms, there seemed to be no advantage to increasing the chain length. This plateau corresponded to a rise in wetting angle from about 55 degrees with 8 carbon atoms to a maximum of 70 degrees above chain lengths of 14 carbon atoms. Buckley (1981) described similar experiments on the lubrication of tungsten single crystals, which showed a decrease of friction coefficient by about a factor of 2 as the number of carbon atoms in the chain increased from 1 to 10. The effects of increasing molecular weight were also observed for pin-on-disk tests of high (100,000–5,000,000) molecular weight polyethylene oxide polymers as well. In addition to petroleum oils, Rabinowicz (1965) listed the following types of liquid lubricants: polyglycols, silicones, chlorofluorocarbons, polyphenyl ethers, phosphate esters, and dibasic esters.

Various ingredients are added to oil base stocks to alter their characteristics and make them more suitable for certain applications. A list of oil additives and their functions is presented in Table 5.13 (Liston, 1992). A specially formulated group of additives for use with base oil is called an additive package, and these packages are adjusted to account for variations in the quality of the crude oil used to produce base oil stock.

Table 5.13 **Additives to lubricating oils**

Additive type	Function
Pour point depressors	High-molecular-weight polymers that inhibit the formation of wax crystals, thereby making the liquid more pourable at lower temperatures
VI improvers	High-molecular-weight polymers that increase the relative viscosity of the oil more at high temperatures than at low temperatures
Defoamers	Silicon polymers at low concentrations, which retard the tendency of oils to foam when agitated
Oxidation inhibitors	Substances added to reduce the oxidation of oils exposed to air, thereby reducing the formation of undesirable compounds and deposits during running
Corrosion inhibitors	Substances added to form protective films on the solid surfaces, which reduces corrosive attack by other species in the oil or the environment
Detergents	Chemically neutralize certain precursors to reduce the formation of deposits
Dispersants	Disperse or suspend potential sludge-forming materials in the oil
Anti-wear additives	Long-chain, boundary lubrication additives to reduce wear
Anti-friction additives	Similar to anti-wear additives in that they enhance contact surface lubricity
Extreme-pressure additives	Form oil-insoluble surface films that help to bear high contact pressures and improve wear and friction as well

Extreme-pressure (EP) additives are used to cause the formation of protective layers on highly loaded bearing surfaces. They consist for the most part of compounds of chlorine, sulfur, and/or phosphorus that react with the surfaces being lubricated (in most cases, ferrous metals). Phosphorus, for example, can react with frictional hot spots on the surface of ferrous bearing surfaces to form low-melting-point phosphide eutectics and thus reduce friction and wear. It is important that additives react very quickly on the bearing surfaces because films removed mechanically during sliding contact must be immediately replenished to maintain stable frictional behaviour. Well-functioning films should be stable when formed, adherent to the surface, and easily sheared.

Friction modifiers and anti-wear additives to oils are proprietary in nature. Tung et al. (1988) described the screening of various compounds and their combinations using a reciprocating laboratory test. High-chromium steel was used as the slider, and several other steels, were used as the counterface alloy. One per cent additions of four different friction modifiers to commercial engine oils of various viscosity grades were used:

FM-1: bis(isoctylphenyl)-dithiophosphates with molybdenum;

FM-2: molybdenum disulfide compound dispersed in an organic carrier;

FM-3: organic sulfur fatty oil; and

FM-4: a sulfur-free organomolybdenum compound.

FM-1 was claimed to be the most effective of the additives tested in reducing friction. In contrast to the other two oils, the friction of the SAE 40 oil seemed little changed by the additives. Additives make it difficult to know exactly how the chemistry of oils changes with exposure to operating conditions. There are tribochemical effects of fuel residues, combustion products, and reaction products due to the wear and corrosion of the materials in contact with the lubricant and effects of temperature and oxidative degradation are of concern. Chlorinated refrigerants, such as CFC-12, have been replaced with less lubricious refrigerants such as HFC-134 (CH_2FCF_3). These new refrigerants require new lubricating additives to reduce friction and wear. One factor that results in high friction is the adhesive transfer of material from one contacting surface to another, leading to self-mated conditions. Metallic materials like iron and steel, copper and brass, aluminium alloys, and titanium transfer relatively easily during sliding contact. As Heinicke (1984) pointed out, fatty amines, fatty alcohols, and fatty acids have been effective for anti-wear and friction-lowering additives, reducing metal transfer by a factor of more than 20,000 times.

With suitable additives and under the proper bearing conditions, even water can be an effective lubricant. Sometimes materials that cannot be lubricated with oils in certain applications, such as in the food-processing industry, might be effectively lubricated with water or with water containing non-toxic additives. For example, Sasaki (1992) has published a comprehensive compilation of the effects of water and water with additives on the friction and wear of ceramics. Sasaki found that silicon carbide exhibits the lowest friction coefficient and seems little affected by sliding speed, in contrast to silicon nitride. Experiments with various glycol additions to water also showed the responsiveness of the friction of silicon nitride couples to water-based lubricant composition.

The pH of water solutions also greatly affected the friction of silicon nitride in other experiments. These examples illustrate that lubrication effectiveness in reducing friction can be a function not only of the sliding speed and geometrical parameters, but also of the composition and pH of the lubricant, factors that are only indirectly incorporated in traditional, mechanically-based bearing design equations through their effects on viscosity. In Sasaki's experiments using water, and in the case of many formulated oils, additives are liquids or species that go into solution, but additives to liquid lubricants and greases can also be used in solid form as dispersants in the fluid. Bhushan and Gupta (1991) discussed the use of various graphite dispersions in liquids. Solid contents can range between one and forty per cent in petroleum oils, and particle sizes can range from 0.5 to 60 μm. Applications for such dispersions of solids in oil range from dies and tooling to engine oils in which the solid dispersant clings to the surface to produce additional anti-wear and friction reduction. Some solid additives are used to provide extra protection of the sliding surfaces should the liquid lubricant fail. Polytetrafluoroethylene (PTFE) in engine oil additive fluids has become popular to reduce engine friction and improve mileage. It can produce lubricity at lower engine temperatures or during starting when full oil films have not yet been developed on the surfaces.

The field of study relating to the effects of the chemical reactions between surfaces and their environment, as they affect friction, lubrication, and wear, is called tribochemistry. Tribochemistry is a very important aspect of lubrication as well as unlubricated friction and wear. In fact, previous discussions of the roles of adhesion, relative humidity, oxidation, film formation, lubrication, and lubricant additives on friction can all be considered part of the wide and complex field of tribochemistry. One of the most comprehensive treatments of tribochemistry is the text by Heinicke (1984), which identifies a number of sub-topics of tribochemistry, including tribodiffusion, tribosorption, tribodesorption, triboreaction, tribooxidation, tribocatalysis, etc. Tribochemistry is a very challenging discipline, since multiple chemical processes can be occurring simultaneously in lubricated tribosystems.

The important role of oxides and other surface films in controlling friction can also affect friction in boundary-lubricated situations. Oxide layer effects were discussed by Komvopoulos et al. (1986). Three metals, oxygen-free high-conductivity Cu, pure Al, and Cr-plate, were oxidized in a furnace to produce various film thicknesses. An additive-free, naphthenic mineral oil was used as the boundary lubricant in self-mated pin-on-disk tests at room temperature in air. A small portion of those authors' friction coefficient results, obtained from plots of their data for 2 N load and at an angular disk rotation speed of 4.5 rad/sec, is summarized in Table 5.14. These data represent only those for the thinnest oxides produced in their experiments, tests that typically ran for about 50 m in sliding distance. Friction in these experiments often exhibited complex behaviour associated with the disruption of the oxides and the incorporation of

Table 5.14 **Effects of oxide scales on boundary-lubricated friction**

Condition/Parameter	Aluminium	Copper	Chromium
Lubricated but not pre-oxidized			
Average μ after 0.1 m sliding	0.45	0.18	0.20
Average μ after 50 m sliding	0.20	0.17	0.14
Lubricated and pre-oxidized			
Oxide layer thicknesses (nm)	4.2	7.0	28.2
Average surface roughness (μm)	0.05	0.1	0.05
μ after 0.1 m sliding	0.12	0.14	0.11
μ after 50 m sliding	0.20	0.14	0.13

debris into the interface. The thicker oxides tended to be more porous than the thinner ones, making them easier to rupture and producing greater quantities of wear debris. Komvopoulos et al. (1986) discussed the surface deformation and wear mechanisms in the interface and suggested several models for the observed behaviour based on microscopy of the contact surfaces.

The roughness of boundary-lubricated surfaces can be altered by the presence of oxides whose growth characteristics change as they thicken, but the surface roughness can also be altered intentionally to modify and reduce friction. For example, Tian et al. (1989) created linear patterns on titanium surfaces subjected to sliding on 52100 bearing steel to study how those regular features affected the ability of certain boundary lubricants to reduce friction. The testing machine slid the steel pin back and forth 30 mm at an average speed of 1.1 cm/sec on the undulated surface at a load of 5 N. The effects of the undulations on the friction of the two metals can be significant, as shown in Table 5.15, but they do not appear to work equally well for all lubricants. Several years earlier, Lancaster and Moorhouse (1985) used photolithography to produce pockets in a range of metal

Table 5.15 **Effect of linear undulations on boundary-lubricated friction of steel on titanium (friction coefficients at steady state)**

Lubricant	μ, Polished	μ, Undulated
Mineral oil	0.60	0.39
Oleic acid	0.47	0.11
Turbo oil	0.48	0.17
Silicone oil	0.46	0.25
Halocarbon oil	0.17	0.17
Methylene iodide	0.18	0.18

substrates for the purpose of creating pockets to hold solid lubricants. In the case of titanium, a difficult to lubricate metal, undulations, coupled with a good choice of lubricant, seem to provide an effective system for lubrication.

Some lubricants can function as solid lubricants over one temperature range, liquids over another, and then become desorbed and cease to function at higher temperatures. Therefore, the conditions of surface contact and the role of the lubricant in separating the surface can change drastically over a range of temperatures. Rabinowicz (1965) illustrated this situation using octadecyl alcohol lubricant between copper sliders. Below 40°C when the lubricant was solid, the friction coefficient was about 0.11, but the system experienced a transition between 40° and 60°C to reach $\mu = 0.33$ when the lubricant became liquefied. Friction remained constant until about 120°C, when another transition to a friction coefficient of about 1.0 occurred as the liquid was ultimately desorbed. The wear rate increased correspondingly at each transition temperature because metal transferred to the opposing surfaces with increasing severity as the friction increased.

Despite the existence of many elegant theories of lubrication and a huge volume of literature on the effects of all manner of experimental parameters on the behaviour of lubricants, how lubricants actually reduce friction is only partially understood. The complexity of additive interactions makes possible many different chemical species in the liquid, the boundary layers, and the surface films. The changing nature of the solid surfaces as the system ages and experiences wear makes reaching a global understanding of lubricated friction very difficult to achieve. Lubricants are collected, filtered, centrifuged, and analysed, and their nature is investigated as functions of temperature, pressure, and exposure to different solid surfaces. Fundamental studies of thin-layer lubrication have been made possible by the recognition that dipping techniques could be used to produce monolayers of lubricant and this technique led to the possibility of investigating lubrication mechanisms on a very fine scale. More recently, there have been advances in molecular-scale measurements of fluid properties. Grannick (1992) and others have described the molecular level behaviour of lubricating films in terms of shear thinning. As the thickness of a film decreases, the friction (shear strength) tends to rise. With the advent of high-speed computers and simulations of interfaces, it has become possible to model the behaviour of molecules in narrow frictional interfaces (Robbins et al., 1993). Such efforts have shown that the structural arrangements of atoms in interfaces change in the vicinity of the solid walls and that the properties of the fluids may be much different adjacent to the boundaries as a result of these changes. Robbins has shown that as the surfaces begin to move, lubricant layers may disorder and then reorder when motion ceases. These fascinating computational results have implications for understanding the nature of boundary lubrication and stick-slip.

Traditional interpretations of boundary lubrication mechanisms have dealt with the orientation of molecules on surfaces. Polar species tend to align with their heads at the surfaces, their tails forming a layer to provide lubricating action. Long-chain fatty acids exemplify this type of behaviour. The shape and side branches of molecules determine whether or not they form dense layers on the surface. Tendencies of liquids to wet surfaces helps their ability to lubricate as well. As fundamental studies of interfacial structure, molecular motions, and tribochemistry are integrated with micro- and macro-mechanics, improvements in lubrication science and technology will emerge.

5.3 Solid lubrication

Additives to liquids can result in the formation of solid films or deposits on contact surfaces to help reduce and control friction. Solid lubricants can be formed in other ways than by reaction within a fluid. For example, Peace (1967) identified eight types of solid lubricant systems based on the method of application and the form of the material:

1. solid lubricant powders;
2. resin-bonded dry-film lubricants;
3. dry film lubricants with inorganic binders;
4. dispersions of solids in a non-volatile carrier;
5. wear-reducing solids (with naturally lubricious surfaces);
6. soft metal films;
7. plastic lubricants; and
8. chemical reaction films (as produced by reactions with lubricant additives, etc.).

Bhushan and Gupta's *Handbook of Tribology* (1991) contains an extensive discussion of coating and surface modification techniques, including those suitable for use with solid lubricants. Additional reviews of solid lubricants may be found in the literature (Clauss, 1972; Lancaster, 1984; Sliney, 1992). More and more solid materials are being found to be lubricious, but the fact that they are lubricious does not constitute a sufficient condition for them to be widely used as solid lubricants. Other factors such as ease of application, thermal stability, adequate persistence on the surface, cost, and chemical compatibility with the surfaces and service environment, are also factors for selection. In addition, some lubricious metals (like Pb) can no longer be used due to increased concerns about their toxicity.

Four primary factors should be considered when selecting and designing solid lubricating films to reduce friction effectively:

1. the structure and composition of the solid lubricant species;
2. the thickness of the solid lubricating film in the given application;
3. the conditions of sliding (contact pressure, velocity, temperature, environment); and
4. the manner by which the solid is resupplied to the surface as sliding or rolling tends to remove it.

The structure and composition of the solid lubricant determine its shear strength, its adhesion characteristics to the substrate, its chemical stability, its durability, and in some cases its tendencies toward anisotropic behaviour. The thickness of the film determines its friction coefficient in much the same way that the Stribeck curve, described earlier, determines the lubrication regime. When films are very thin, asperities can penetrate and disrupt, so that thinner films may work if the surfaces are polished extremely flat. When films become too thick, they behave more like bulk solids. For example, silver is an effective solid lubricant when in thin film form, but the friction coefficient of bulk silver rubbing on the same material can be more than ten times higher.

The method of surface preparation prior to the application of solid lubricants must be carefully considered if the full benefits of the solid lubricant are to be achieved. For steels and stainless steels, surface preparation may involve phosphate coating, grit blasting, or grinding, while for aluminium alloys, anodization treatments may be applied. Titanium alloys may require grit blasting and/or chemical etching. One method of applying solid lubricants is to incorporate them within a synthetic resin binder and to paint or spray them onto a surface. Some of the formulations cure in air, but others require oven drying and curing. According to Lancaster (1984), several common constituents of bonded-film lubricants are MoS_2, WS_2, graphite, PTFE, pthalocyanine, CaF_2/BaF_2, Pb, PbO, PbS, Sb_2O_3, Au, Ag, and In. Sonntag (1960) has tabulated data for the static and kinetic friction coefficient of solid lubricants on metals, and indicated whether or not they exhibited tendencies for stick-slip. A selection of these data is provided in Table 5.16. Since the conditions of use vary greatly and may differ significantly from Sonntag's testing conditions, these values are only provided as an example of relative

Table 5.16 **Friction coefficients for steel lubricated by solid lubricants**

Lubricant	μ_s	μ_k	S-S[a]
None (steel-on-steel)	0.40–0.80	0.40	
Molybdenum disulfide	0.05–0.11	0.05–0.093	N
Tungsten disulfide	0.098	0.09	N
Selenium disulfide, titanium	—	0.25	Y
Mica, talc	—	0.25	N
Graphite	—	0.25	N
Boron nitride (hexagonal)	—	0.25	Y
Vermiculite	0.167	0.160	N
Beeswax (at 60–63°C)	0.055	0.05	N
Paraffin (at 47–77°C)	0.112	0.104	N
Calcium stearate (157–163°C)	0.113	0.107	N
Carnauba wax (83–86°C)	0.169	0.143	N
Sodium stearate (198–210°C)	0.192	0.164	Y
Lithium 12-hydroxystearate (210–215°C)	0.218	0.211	N

[a] S-S, tendency for stick-slip; Y, yes; N, no.

differences in the solid lubricating behaviour of various materials at room temperature.

Solid lubricants are also used at elevated temperatures or in high vacuum applications where most liquid lubricants would volatilize, oxidize, or otherwise become unstable. Table 5.17 lists the friction coefficients of several candidate solid lubricants obtained under the same high-temperature sliding conditions (7.7 kgf and 7.6 mm/sec on steel at 704°C), as reported by Peterson et al. (1969). A later compilation of high-temperature solid lubricant friction coefficients was produced by Allam (1991). Table 5.18 provides room-temperature friction values of compressed powder pellets sliding on stainless steel from a compilation by Clauss (1972).

Many of the most important solid lubricating materials exhibit what has been called lamellar behaviour. That is, there tend to be weak shear planes within the structure of the material that can yield preferentially to reduce friction, if they are properly aligned to the sliding direction. Many of the compounds in Table 5.18 form hexagonal crystal

Table 5.17 Kinetic friction coefficients for several oxides at 704°C

Lubricating solid	μ
PbO	0.12
B_2O_3	0.14
MoO_3	0.20
Co_2O_3	0.28
Cu_2O	0.44
SnO	0.42
TiO_2	0.50
MnO_2	0.41
K_2MoO_4	0.20
Na_2WO_4	0.17

Table 5.18 **Properties and friction coefficients characteristic of certain compounds**

Class	Compound	Crystal structure	μ
Disulfides	MoS_2	Hexagonal	0.21
	WS_2	Hexagonal	0.142
	NbS_2	Hexagonal	0.098
	TaS_2	Hexagonal	0.033
Diselenides	$MoSe_2$	Hexagonal	0.178
	WSe_2	Hexagonal	0.13
	$NbSe_2$	Hexagonal	0.12
	$TaSe_2$	Hexagonal	0.084
Ditellurides	$MoTe_2$	Hexagonal	0.20
	WTe_2	Orthorhombic	0.38
	$NbTe_2$	Trigonal	0.70
	$TaTe_2$	Trigonal	0.53
Graphite	C	Hexagonal	0.14

structures in which the shear strength is lowest parallel to the basal planes. In the case of molybdenum disulfide, there are weak van der Waals' bonds between covalently bonded Mo–S layers. Moisture and air tend to reduce the effectiveness of MoS_2 as a solid lubricant, since they penetrate these layers and raise their shear stresses. On the other hand, graphite is observed to be more lubricious in moist environments, since interlayer species reduce the shear strength. Therefore, molybdenum disulfide is very effective in low vacuum (space) applications, but graphite is more effective in moist environments. Table 5.19 illustrates some of these effects, but it should be noted that solid lubricants used in powdered form might produce different effects on friction than the same compositions applied by other methods.

The friction of solid lubricants is sometimes modelled by assuming that the friction force F in a given sliding system is

Table 5.19 **Effects of moisture on the friction coefficients of various solid lubricants in air of various relative humidity**

Solid lubricant	μ_k, dry air, RH <6%	μ_k, moist air, RH = 85% (after sliding in dry air)	μ_k, dry air (after sliding in moist air)
Molybdenum disulfide powder	0.06	0.20	0.06
Molybdenum disulfide bonded film on disk	0.09	0.22	0.09
Molybdenum disulfide bonded film on both slider and disk	0.26	0.34	0.31
Graphite powder	0.06–0.10[a]	0.16	0.19

[a] Initial value before film failure.

determined by the shear strength τ of the interfacial medium:

$$\mathbf{F} = \tau A \qquad [5.55]$$

where A is the contact area over which shear force **F** is acting. As discussed earlier in regard to friction modelling, it has been found by Bridgman (1935) that τ is a function of the contact pressure p. Thus,

$$\tau = \tau_0 + \alpha p \qquad [5.56]$$

and the pressure coefficient α determines the change in the saturation shear stress with pressure. Bednar et al. (1993) have re-examined the pressure dependence of the yield strength and, using anvil experiments, determined the effective friction coefficient μ_{eff} of metals as a function of applied pressure p and the saturation yield stress (represented as τ in eq. [5.55]. Thus,

$$\mu_{\text{eff}} = \tau / p \qquad [5.57]$$

Experimentally determined values for τ_0 and α of several

Table 5.20 **Dependence of saturation shear strength and friction of metals on the applied pressure**

Metal	α	τ_0 (MPa)	μ_{eff} (at $p = 1$ MPa)
Fe	0.075	173.82	0.246
Cu	0.049	107.61	0.160
Ag[a]	0.036	109.04	0.144
Au	0.029	91.82	0.123
Al	0.035	47.56	0.082
Sn[a]	0.012	12.30	0.024
In[a]	0.006	5.65	0.011

[a] Commonly used solid lubricants.

metals are listed in Table 5.20. Of the three metals listed as solid lubricants, silver and tin are used more than indium. Interestingly, the frictional response of silver is quite suitable for solid lubrication because there is no significant difference between μ_s and μ_k, leading to very smooth sliding with an absence of stick-slip behaviour.

Graphite and molybdenum disulfide are among the most commonly used solid lubricants, and it is worthwhile to consider their frictional behaviour specifically. Winer (1967) compiled an extensive review of molybdenum disulfide as a solid lubricant that was published in 1967, and Fleischauer and Bauer (1987) reviewed the chemistry and structure of sputtered MoS_2 films. Clauss (1972) has reviewed lubrication by graphite. Both these materials are anisotropic in properties due to their hexagonal crystal structures, which contain wide separations between the basal planes. When properly oriented, for example, by running in the surface to produce easy-shear platelet orientations, friction coefficients for MoS_2 can be as low as 0.02–0.1. Graphite typically exhibits friction coefficients of 0.10–0.15 in air. Fluorination of graphite to produce substoichiometric graphite fluoride

CF_x $(0.3 \leq x \leq 1.1)$ increases the spacing between the basal planes, making it a good candidate for future lubricants. Another method to spread the basal planes and enhance graphite lubrication is by intercalation. Intercalation involves the insertion of atoms between basal planes. This process can result in significant enhancements of film life as well as frictional performance.

As discussed by Peace (1967), the maximum use temperatures for graphite and molybdenum disulfide depend on other factors than temperature alone. These include relative humidity, oxygen concentration in the environment, and whether the material is in powdered or monolithic form. In furnace oxidation experiments, graphite powder begins to oxidize significantly at about 585°C, compared with 298°C for molybdenum disulfide powder. Fusaro (1978) found that oxidation causes molybdenum disulfide films to blister and fail. This is explained by the tendency of molybdenum disulfide to form oxides and sulfides of various stoichiometry in air. In the case of graphite, the oxidation is highly anisotropic, but the rates of oxidation are slower than for molybdenum disulfide.

Bisson and Anderson (1964) prepared an extensive review of solid lubricant properties, including graphite, molybdenum disulfide, and molybdenum trioxide. For example, the friction coefficients of various MoS_2 and MoO_3 films on steel surfaces reaction to increasing sliding velocities is stunning. Clearly, MoO_3 is a very poor lubricant. As the temperature increases in air, molybdenum disulfide undergoes changes in colour and rate of oxidation. Table 5.21 summarizes these changes, as discussed by Bisson and Anderson. Molybdenum disulfide and graphite can each be used as solid lubricants, but attempts have been made to determine whether mixing them together would provide synergistic effects. Gardos (1987), for example, reviewed the use of graphite as an oxygen scavenger to help molybdenum disulfide films retain

Table 5.21 **Transformations in molybdenum disulfide as temperature rises**

Temperature range (°F)	Temperature range (°C)	Behaviour
Up to 750	Up to 400	No detectable oxidation rate
750–800	400–427	Thin oxide film forms
800–850	427–454	Slow, but appreciable oxidation
850–900	454–482	Yellowish white MoO forms
Over 900	Over 482	Rapid oxidation

low friction characteristics. Some limited advantages in enhancing the stability and wear resistance of the microcrystalline molybdenum disulfide films in air were reported. About the same time, Bartz et al. (1986) in Germany studied the friction of bonded films containing various combinations of graphite, molybdenum disulfide, and antimony thioantimonate [$Sb(SbS_4)$]. Using a block-on-ring apparatus, after sandblasting the 100CrMn6 steel ring, they applied bonded films to it, but left the 90MnCrV8 steel block untreated. A method of assessing the effectiveness of the blends was to measure the stable, post-running-in friction coefficient. Table 5.22 lists values of μ for several combinations of lubricants.

Briscoe (1992) has reviewed the mechanisms of organic polymer friction, stating that two non-interacting contributions, adhesion and plowing, can be used to model behaviour. In this treatment, frictional energy is dissipated by an interface zone (adhesive) and a subsurface zone (deformation and/or plowing). In the latter zone, behaviour in polymers may be viscoelastic, plastic, or brittle. The friction coefficient

Table 5.22 **Steady-state friction coefficients for solid lubricant combinations**

Film composition	μ, Steady state
Graphite alone	Unstable μ
MoS_2 alone	0.05
Graphite + MoS_2 (about 1:2 wt% ratio)	0.01–0.02
Graphite + $Sb(SbS_4)$ (about 3:4 wt% ratio)	Unstable μ
MoS_2 + $Sb(SbS_4)$ (about 4:5:1 wt% ratio)	0.1–0.03
Graphite + MoS_2 + $Sb(SbS_4)$	0.04–0.05

could be derived from geometric arguments to produce the form

$$\mu = \frac{2}{\pi} \tan \theta \qquad [5.58]$$

where the angle θ was associated with the roughness of the surface. Briscoe found that if PTFE behaved in a more brittle fashion, as it did after irradiation by gamma rays, the same expression could be similarly written:

$$\mu = \frac{x}{\pi} \tan \theta \qquad [5.59]$$

where the value of x, the slope of the dependence of friction on tan θ, varied from 0 to 2 depending on the degree of embrittlement (i.e., the extent of plastic flow). When tan θ exceeded approximately 2.3, irrespective of x, the PTFE began to exhibit chip-forming characteristics rather than flow. Like other materials, the shear stress of PTFE was seen to vary with contact pressure.

Since PTFE has a relatively low hardness, various additives are mixed with it to improve its wear resistance. Table 5.23 shows the effects of certain additives on the friction coefficient and wear of PTFE sliding on steel at 0.01 m/sec. Comparing the first and last row of data shows how it is possible to increase wear resistance by more than three orders of magnitude while

Table 5.23 **Effect of additives on the friction of blended PTFE**

Material composition	Wear rate improvement[a]	μ_k
Unfilled PTFE	1	0.10
15 wt% graphite	588	0.12
15 wt% glass fibre	2857	0.09
12.5 wt% glass fibre and 12.5 wt% MoS_2	3333	0.09
55 wt% bronze and 5 wt% MoS_2	4000	0.13

[a]Ratio of the wear rate of unfilled PTFE to that of the given material.

raising the sliding friction coefficient of the material by at most about 0.03. Erdemir (1994) reviewed some of the important mechanisms responsible for the lubricating action of solid films on ceramics such as silicon carbide, silicon nitride, and aluminium oxide. He stated that solid lubrication may be the only option available to help lubricate ceramics in severe environments, but noted that like other types of lubricants, solid lubricants suffer from finite lifetimes. He discussed the use of boric oxide (B_2O_3) and its product with water, and boric acid (H_2BO_3) in particular. Boric acid resembles other lubricants with layered structures, and produces favourable friction reductions under some circumstances. However, when the temperature rises above about 170°C, boric acid decomposes to boric oxide and loses its layered structure.

5.4 Tribology of rolling

Rolling is a process that cannot be conducted without friction as friction is needed to draw the work piece into the roll gap and to deform it. The minimum value of friction is twice that needed for continuous rolling. The effects of friction are connected to the geometric description of the process known as the L/h ratio, where L is the projected length of the arc of contact and h is the mean strip thickness. At $L/h > 2$, deformation is homogeneous and the limiting strip thickness may be reached. At $L/h < 2$, there is an inhomogeneity and at $L/h < 1$, sticking friction occurs, so lubricants are applied to reduce friction and wear. In cold rolling, lubricants are used to reduce friction, although a minimum amount is required. Surface finish requirements are friction dominated. In hot rolling, lubricants are used to control adhesion between material and roll. Lubricants may be oil or water based, and extreme pressure additives are used where there is a mixed-film lubricating mechanism. The most commonly used lubricants in rolling are shown in Table 5.24.

Table 5.24 **Commonly used lubricants and typical μ (friction coefficient) values in cold and hot rolling**

Material	Hot rolling - lubricant	Hot rolling - μ	Cold rolling - lubricant	Cold rolling - μ
Steel	Water Emulsion of fat+EP additive Fat (ester)+EP additive +water	Sticking 0.4 0.3	3–6% emulsion of palm oil	0.01–0.03
Al and Mg alloys	Emulsion, 2–15% of mineral oil	0.4	Mineral oil with 1–5% fatty acid	0.01–0.03
Cu and Cu alloys	Emulsion, 2–8% of mineral oil	0.3	2–10% concentration of mineral oil with fat	0.01–0.03
Ti alloys	Fat+water	Sticking	Esters or soap Castor oil Compounded mineral oil	0.2 0.2 0.2
Refractory metals	Dry	0.3	Mineral oil with boundary and EP additives	0.01–0.03

5.5 Tribology of drawing

Drawing is a process where the size of a work piece is reduced by pulling through a constriction. No friction is required between work piece and die in wire drawing and tube drawing on a fixed plug. Moderate friction is needed on the plug for drawing tubes and frictional of a bar is beneficial when drawing a bar. Lubricants are applied in drawing to reduce friction, wear, and temperature and the method of application is critical in drawing. Drawing without a lubricant results in material pick-up. Dry drawing is conducted with soap, whereas wet drawing is conducted with viscous oils or aqueous emulsions. Most practical drawing is conducted under mixed film regimes where lubricants, soaps and extreme pressure additives can yield the best results at various stages of drawing. The most commonly used lubricants in drawing are shown in Table 5.25.

Table 5.25 Commonly used lubricants and typical μ (friction coefficient) values in wire and tube drawing

Material	Wire drawing - lubricant	Wire drawing - μ	Tube drawing - lubricant	Tube drawing - μ
Steel	Mineral oil + fat + EP additive Phosphate + emulsion	0.07 0.1	Phosphate + soap	0.05
Al and Mg alloys	Mineral oil + fatty derivatives	0.03–0.15 (mixed film)	Soap	0.07
Cu and Cu alloys	Mineral oil + fatty derivatives	0.03–0.15 (mixed film)	Soap film	0.05
Ti alloys	Fluoride phosphate + soap	0.1	Metal + soap	0.07
Refractory metals	Copper + mineral oil	0.1	Copper + mineral oil	0.1

5.6 Tribology of extrusion

Extrusion refers to the metalworking process that pushes metal through a constriction of known geometry and friction is generally unnecessary and undesirable. Friction on the die increases extrusion pressure and impairs the homogeneity of deformation. Friction also contributes to heat generation and limits attainable reductions and speeds in hot extrusion. Extrusion can be fully lubricated or unlubricated. Unlubricated extrusion is essential for extruding tubes, hot extrusion of aluminium alloys with flat dies, and non-isothermal extrusion of copper alloys. The most commonly used lubricants in extrusion are shown in Table 5.26.

5.7 Tribology of forging

In simple open die forging operations, friction induces in-homogeneity of deformation and increases forging pressures.

Table 5.26 Commonly used lubricants and typical μ (friction coefficient) values used in extrusion of metals

Material	Hot extrusion – lubricant	Hot extrusion – μ	Cold extrusion – lubricant	Cold extrusion – μ
Steel	Graphite	0.2	NA	NA
Al and Mg alloys	None	Sticking friction	Lanolin	0.07
Cu and Cu alloys	Graphite	0.2	Castor oil	0.03
Ti alloys	Graphite	0.2	NA	NA
Refractory metals	Glass coating plus graphite on die	0.05	NA	NA

Lubrication is beneficial in reducing die wear and other duties such as reducing local forging pressure. Lubrication in cold forging relies on compounded oils and semi-solids such as fats, soaps, and waxes. For severe working, MoS_2 may be used. In hot forging applications, lubricants are based on oil based graphitic solutions, or graphite free solutions. The most commonly used lubricants in forging are shown in Table 5.27.

5.8 Tribology of sheet metalworking

Friction does not affect the processes of shearing, blanking, and punching itself, but lubrication is used to reduce die

Table 5.27 **Commonly used lubricants and typical μ (friction coefficient) values used in forging operations**

Material	Hot forging – lubricant	Hot forging – μ	Cold forging – lubricant	Cold forging – μ
Steel	Soap Graphite in water Salt solution	0.3 0.2 0.2	Lime + oil Copper + oil Phosphate + soap	0.1 0.1 0.05
Al and Mg alloys	Graphite in water	0.3	Lanolin Phosphate + soap	0.07 0.05
Cu and Cu alloys	Graphite in water	0.15	Fat: wax (lanolin) Zinc stearate (soap) Graphite or MoS_2 in grease	0.07 0.05 0.07
Ti alloys	MoS_2	0.2	Zinc Fluoride-phosphate + soap	0.1 0.05
Refractory metals	Glass + graphite	0.05	NA	NA

wear. Sheet metalworking dies tend to wear by adhesive, abrasive, fatigue and chemical wear mechanisms. Bending places minor demands on lubricants, whereas in spinning and flow turning, high pressures require good lubrication. Stretching processes require good lubrication as friction governs the shape of the component to be produced. In deep drawing, lubricants dominate the level of LDR values requiring high viscosity lubricants. The most commonly used lubricants in sheet metalworking are shown in Table 5.28.

Table 5.28 Commonly used lubricants and typical μ (friction coefficient) values used in sheet metalworking operations

Material	Shearing - lubricant	Bending - lubricant	Press working - Lubricant	Press working - μ
Steel	Pickle oil, emulsion of mineral oil and EP additives	Pickle oil, emulsion of mineral oil and EP additives	Phosphate + soap, wax, graphite in grease, metal + emulsion	0.05
Al and Mg alloys	Emulsion of mineral oil and EP additives	Emulsion of mineral oil and EP additives	Soap or lanolin, polymer, graphite coating	0.05
Cu and Cu alloys	Soap solution, mineral oil + fat, emulsion of mineral oil and fat	Soap solution, mineral oil + fat, emulsion of mineral oil and fat	Tallow, pigmented tallow, soap	0.05–0.1
Ti alloys	Mineral oil + EP additives	Mineral oil + EP additives	Wax, polymer, fluoride phosphate + soap	0.05–0.07
Refractory metals	Mineral oil + EP additives	Mineral oil + EP additives	MoS_2 or graphite, Al-Fe-Bronze dies with wax	0.2 0.07

5.9 Conclusions

This chapter has focused on tribology in manufacturing processes from the viewpoint of understanding the fundamentals of sliding friction in those processes and the use of lubricants to control friction in manufacturing processes such as machining, drawing, rolling, extrusion, abrasive processes and processing at the micro and nanoscales. The chapter also provides data on the type of lubricants used in manufacturing practice and how the lubricant controls friction to achieve a variety of different effects at various length scales. It should be noted that there is a lack of genuine data of how lubricants affect the tribology of manufacturing at the micro and nanoscales and this must surely be the challenge for future engineers working in this exciting field.

5.10 References

Akhmatov, A. S. (1939), 'Some items in the investigation of the external friction of solids, Trudy Stankina'; cited by I. V. Kragelski (1965) in *Friction and Wear*, Butterworths, London, p. 159.

Allam, I. M. (1991), 'Solid lubricants for applications at elevated temperatures', *J. Mater. Sci.* **26**: 3977–84.

Bartenev, G. M. and Lavrentev, V. V. (1981), *Friction and Wear of Polymers*, Elsevier, New York, pp. 53–61.

Bartz, W. J., Holinski, R. and Xu, J. (1986), 'Wear life and frictional behavior of bonded solid lubricants', *Lubr. Eng.* **42**(12): 762–9.

Bednar, M. S., Cai, B. C. and Kuhlmann-Wilsdorf, D. (1993), 'Pressure and structure dependence of solid lubrication', *Lub. Eng.* **49**(10): 741–9.

Belak, J. and Stowers, J. F. (1992), 'The indentation and scraping of a metal surface: a molecular dynamics study', in *Fundamentals of Friction: Macroscopic and Microscopic Processes*, eds. H. M. Pollock and I. L. Singer, Kluwer, Dordrecht, The Netherlands, pp. 511–20.

Bhushan, B. and Gupta, B. K. (1991), *Handbook of Tribology*, McGraw-Hill, New York, pp. 5–11 and 5–12.

Bisson, E. E. and Anderson, W. J. (1964), *Advanced Bearing Technology*, NASA SP-38, Washington, DC.

Black, P. H. (1961), *Theory of Metal Cutting*, McGraw-Hill, New York, Chapter 5, pp. 45–72.

Blok, H. (1937), *General Discussion on Lubrication*, Inst. of Mechanical Engineers, Vol. 2, p. 222.

Blok, H. (1963), 'The flash temperature concept', *Wear*, 6: 483–94.

Boehringer, R. H. (1992), 'Grease', in *ASM Handbook*, 10th edn, Vol. 18, *Friction, Lubrication, and Wear Technology*, ASM International, Materials Park, OH, pp. 123–31.

Bowden, F. P., and Tabor, D. (1986), *The Friction and Lubrication of Solids*, Clarendon Press, Oxford.

Bridgman, P. W. (1931), *The Physics of High Pressure*, Macmillan Press, New York.

Bridgman, P. W. (1935), 'Effects of high shearing stress combined with high hydrostatic pressure', *Phys. Rev.* 48: 825–47.

Briscoe, B. J. (1992), 'Friction of organic polymers', in *Fundamentals of Friction: Macroscopic and Microscopic Processes*, eds. I. L. Singer and H. M. Pollock, Kluwer, Dordrecht, The Netherlands, pp. 167–82.

Briscoe, B. J. and Stolarski, T. A. (1993), 'Friction', Chapter 3 in *Characterization of Tribological Materials*, ed. W. A. Glaeser, Butterworth Heinemann, Boston, pp. 48–51.

Buckley, D. F. (1981), *Surface Effects in Adhesion, Friction, Wear, and Lubrication*, Elsevier, New York, Chapter 5, pp. 245–313.

Burton, R. A. (ed.) (1980), *Thermal Deformation in Frictionally-Heated Systems*, Elsevier, Lausanne, Switzerland, p. 290.

Campbell, W. E. (1940), Remarks printed in *Proc. M.I.T. Conference on Friction and Surface Finish*, MIT Press, Cambridge, MA, p. 197.

Carson, G., Hu, H.-W. and Grannick, S. (1992), 'Molecular tribology of fluid lubrication: Shear thinning', *Tribal. Trans.* **35**(3): 405–10.

Challen, J. M. and Oxley, P. L. B. (1979), 'An explanation of the different regimes of friction and wear using asperity deformation models', *Wear* **53**: 229–43.

Clauss, F. J. (1972), *Solid Lubricants and Self-Lubricating Solids*, Academic Press, New York, pp. 114–15.

Cowan, R. S. and Winer, W. O. (1992), 'Frictional heating calculations', in *ASM Hand-book*, 10th edn, Vol. 18, *Friction, Lubrication, and Wear Technology*, ASM International, Materials Park, OH, pp. 39–44.

Dow, T. A. and Stockwell, R. D. (1977), 'Experimental verification of thermoelastic instabilities in sliding contact', *J. Lubrication Technol.* **99**(3): 359.

Enthoven, J. C., Cann, P. M. and Spikes, H. A. (1993), 'Temperature and scuffing', *Tribol. Trans.* **36**(2): 258–66.

Erdemir, A. (1994), 'A review of the lubrication of ceramics with thin solid films', in *Friction and Wear of Ceramics*, ed. S. Jahanmir, Marcel Dekker, New York, pp. 119–62.

Ferrante, J., Bozzolo, G. H., Finley, C. W. and Banerjea, A. (1988), 'Interfacial adhesion: Theory and experiment', in *Adhesion in Solids*, eds. D. M. Mattox, J. E. E. Baglin, R. J. Gottschall, and C. D. Batich, Materials Research Society, Pittsburgh, PA, pp. 3–16.

Fleischauer, P. D. and Bauer, R. (1987), 'Chemical and structural effects on the lubrication properties of sputtered MoS_2 films', *Tribal. Trans.* **31**(2): 239–50.

Friberg, S. E., Ward, A. J., Gunsel, S. and Lockwood, F. E. (1989). Lyotropic liquid crystals in lubrication, in Ref. 51, pp. 101–111.

Fusaro, R. L. (1978), *Lubrication and Failure Mechanisms of Molybdenum Disulfide Films, I – Effect of Atmosphere*, National Aeronautical and Space Administration special pub., NASA TP-1343.

Gardos, M. N. (1987), 'The synergistic effects of graphite on the friction and wear of MoS_2 films in air', *Tribol. Trans.* **31**(2): 214–27.

Glaeser, W. A. and Dufrane, K. F. (1978), 'New design methods for boundary lubricated sleeve bearings', *Machine Design*, **6**: 207–13.

Grannick, S. (1992), 'Molecular tribology of fluids', in *Fundamentals of Friction*, ed. I. L. Singer and H. M. Pollock, Kluwer, Dordrecht, The Netherlands, pp. 387–96.

Greenwood, J. A. (1992), 'Problems with surface roughness', in *Fundamentals of Friction: Macroscopic and Microscopic Processes*, ed. I. L. Singer and H. M. Pollock, Kluwer, Dordrecht, The Netherlands, pp. 57–76.

Greenwood, J. A. and Williamson, J. B. P. (1966), 'Contact of nominally flat surfaces', *Proc. Royal Soc. London A*, **295**: 300–19.

Griffioen, J. A., Bair, S. and Winer, W. O. (1985), 'Infrared surface temperature in a sliding ceramic-ceramic contact', in *Mechanisms of Surface Distress*, ed. D. Dowson et al., Butterworths, London, pp. 238–45.

Hall, A. S., Holowenko, A. R. and Laughlin, H. G. (1961), 'Lubrication and bearing design', in *Machine Design*, Schaum's Outline Series, McGraw-Hill, New York, p. 279.

Harris, J. H. (1967), 'Lubricating greases', in *Lubricants and Lubrication*, ed. E. R. Braithwaite, Elsevier, Amsterdam, pp. 197–268.

Heinicke, G. (1984), *Tribochemistry*, Carl Hanser Verlag, Munich, p. 446.

Hirst, W. and Hollander, A. E. (1974), *Proc. Royal Soc. London A* **233**: 379.

Hokkirigawa, K. and Kato, K. (1988), 'An experimental and theoretical investigation of ploughing, cutting and wedge formation during abrasive wear', *Tribol. Int.* **21**(1): 51–7.

Hutchings, I. M. (1992), *Tribology – Friction and Wear of Engineering Materials*, CRC Press, Boca Raton, FL, p. 65.

Israelachvili, J. N. (1992), 'Adhesion, friction, and lubrication of molecularly smooth surfaces', in H. M. Pollock and I. L. Singer, eds., *Fundamentals of Friction: Macroscopic and Microscopic Processes*, Kluwer, Dordrecht, The Netherlands, pp. 351–81.

Jaeger, J. C. (1942), *J. Proc. Royal Soc. N. South Wales* **76**: 203.

Klaus, E. E. and Tewksbury, E. J. (1984), 'Liquid lubricants', in *Handbook of Lubrication (Theory and Practice of Tribology)*, Vol. II, ed. E. R. Booser, CRC Press, Boca Raton, FL, pp. 229–54.

Komvopoulos, K. Saka, N. and Suh, N. P. (1986), 'The significance of oxide layers in boundary lubrication', *J. Tribol.* **108**: 502–13.

Kosterin, J. I. and Kragelski, I. V. (1962), 'Rheological phenomena in dry friction', *Wear* 5: 190–7.

Kragelski, I. V. (1965), *Friction and Wear*, Butterworths, London, p. 200.

Kragelskii, I. V. Dobychin, M. N. and Kombalov, V. S. (1982), *Friction and Wear Calculation Methods*, Pergamon Press, Oxford, pp. 178–80.

Kudinov, V. A. and Tolstoy, D. M. (1986), 'Friction and oscillations', in *Tribology Handbook*, ed. I. V. Kragelski, Mir, Moscow, p. 122.

Kuhlmann-Wilsdorf, D. (1987), 'Demystifying flash temperatures I. Analytical expressions based on a simple model', *Mater. Sci. Eng.* **93**: 107–17.

Lancaster, J. K. (1984), 'Solid lubricants', in *CRC Handbook of Lubrication*, ed. E. R. Booser, CRC Press, Boca Raton, FL, pp. 269–90.

Lancaster, J. K. and Moorhouse, P. (1985), 'Etched pocket bearing materials', *Tribol. Int.* **18**(3): 139–49.

Landman, U., Luetke, W. D., Burnham, N. A. and Colton, R. J. (1990), 'Atomistic mechanisms and dynamics of adhesion, nanoindentation, and fracture', *Science* **248**: 454–61.

Lee, H. S., Winoto, S. H., Winer, W. O., Chiu, M. and Friberg, S. E. (1989), 'Film thickness and frictional behavior of some liquid crystals in concentrated point contacts', in *Tribology and the Liquid–Crystalline State*, American Chemical Society, Washington, DC, pp. 113–21.

Liston, T. V. (1992), 'Engine lubricant additives – what they are and how they function', *Lubr. Eng.* **48**(5): 389–97.

McClelland, G. M., Mate, C. M., Erlandsson, R. and Chiang, S. (1987), *Phys. Rev. Lett.* **59**: 1942.

McClelland, G. M., Mate, C. M., Erlandsson, R. and Chiang, S. (1988), 'Direct observation of friction at the atomic scale', in *Adhesion in Solids*, eds. D. M. Mattox, J. E. E. Baglin, R. J. Gottschall, and C. D. Batich, Materials Research Society, Pittsburgh, PA, pp. 81–6.

McCool, J. (1986), 'Comparison of models for the contact of rough surfaces', *Wear* **107**: 37–60.

Miyoshi, K. and Buckley, D. H. (1981), 'Anisotropic tribological properties of silicon carbide', *Proc. Wear of Materials*, ASME, New York, pp. 502–9.

Moore, D. F. (1975), *Principles and Applications of Tribology*, Pergamon Press, Oxford, p. 152.

Mulhearn, T. O. and Samuels, L. E. (1962), *Wear* **5**: 478.

Peace, J. B. (1967), 'Solid lubricants', Chapter 2 in *Lubrication and Lubricants*, ed. E. R. Braithwaite, Elsevier, Amsterdam, pp. 67–118.

Peterson, M. B., Murray, S. F. and Florek, J. J. (1969), 'Consideration of lubricants for temperatures above 1000°F', *ASLE Trans.* **2**: 225–34.

Pollock, H. M. and Singer, I. L. eds. (1992), *Fundamentals of Friction: Macroscopic and Microscopic Processes*, Kluwer, Dordrecht, The Netherlands, p. 621.

Quinn, T. F. J. and Winer, W. O. (1987), 'An experimental study of the "hot spots" occurring during the oxidational wear of tool steel on sapphire', *J. Tribol.* **109**(2): 315–20.

Rabinowicz, E. (1956), 'Stick and slip', *Sci. Am.* **195**(5): 109–18.

Rabinowicz, E. (1965), *Friction and Wear of Materials*, John Wiley and Sons, New York.

Rabinowicz, E. (1992), 'Friction coefficients of noble metals over a range of loads', *Wear* **159**: 89–94.

Raimondi, A. A. (1968), 'Analysis and design of sliding bearings', Chapter 5 in *Standard Handbook of Lubrication Engineering*, McGraw-Hill, New York.

Ramesh, K. T. and Clifton, R. J. (1987), 'A pressure-shear plate impact experiment for studying the rheology of lubricants at high pressures and high shear rates', *J. Tribol.* **109**: 215–22.

Robbins, M. O. and Thompson, P. A. (1991), 'The critical velocity of stick-slip motion', *Science* **253**: 916.

Robbins, M. O. Thompson, P. A. and Grest, G. S. (1993), 'Simulations of nanometer-thick lubricating films', *Mater. Res. Soc. Bull.* **XVIII**(5): 45–9.

Sanchez-Rubio, M., Heredia-Veloz, A., Puig, J. E. and Gonzalez-Lozano, S. (1992), 'A better viscosity-temperature relationship for petroleum products', *Lubr. Eng.* **48**(10): 821–6.

Sasaki, S. (1992), 'Effects of water-soluble additives on friction and wear of ceramics under lubrication with water', in *Effects of Environment on the Friction and Wear of Ceramics*, Bulletin of the Mechanical Engineering Laboratory, Japan, No. 58, pp. 32–53.

Sikorski, M. E. (1964), 'The adhesion of metals and factors that influence it', in *Mechanisms of Solid Friction*, eds. P. J. Bryant, L. Lavik, and G. Salomon, Elsevier, Amsterdam, pp. 144–62.

Sliney, H. E. (1992), 'Solid lubricants', in *ASM Handbook*, 10th edn, Vol. 18, *Friction, Lubrication, and Wear Technology*, ASM International, Materials Park, OH, pp. 113–22.

Song, J. F. and Vorburger, T. V. (1992), 'Surface texture', in *ASM Handbook*, 10th edn, Vol. 18, *Friction, Lubrication, and Wear Technology*, ASM International, Materials Park, OH, pp. 334–45.

Sonntag, A. (1960), 'Lubrication by solids as a design parameter', *Electro-Technology* **66**: 108–15.

Suh, N. P. (1986), *Tribophysics*, Prentice Hall, Englewood Cliffs, NJ, pp. 416–24.

Thompson, P. A. and Robbins, M. O. (1990), 'Origin of stick-slip motion in boundary lubrication', *Science*, **250**: 792–4.

Tian, H., Saka, N. and Suh, N. P. (1989), 'Boundary lubrication studies on undulated titanium surfaces', *Trib. Trans.* **32**(3): 289–96.

Tung, C.-Y., Hsieh, S. K., Huang, G. S. and Kuo, L. (1988), 'Determination of friction-reducing and antiwear characteristics of lubricating engine oils compounded with friction modifiers', *Lubr. Eng.* **44**(10): 856–65.

Wheeler, D. R. (1975), *The Effect of Adsorbed Chlorine and Oxygen on the Shear Strength of Iron and Copper Junctions*, NASA TN D-7894.

Whitehouse, D. J. and Archard, J. F. (1970), 'The properties of random surfaces of significance in their contact', *Proc. Royal Soc. London A* **316**: 97–121.

Wills, J. G. (1980), *Lubrication Fundamentals*, Marcel Dekker, New York.

Winer, W. O. (1967), 'Molybdenum disulfide as a lubricant: A review of the fundamental knowledge', *Wear* **10**: 422–51.

Young, W. C. (1989), *Roark's Formulas for Stress and Strain*, 6th edn, McGraw-Hill, New York.

Zisman, W. A. (1959), 'Durability and wettability properties of monomolecular films on solids', in *Friction and Wear*, ed. R. Davies, Elsevier, Amsterdam, pp. 110–48.

6

Bio and medical tribology

S. Affatato and F. Traina, Istituto Ortopedico Rizzoli, Bologna, Italy

Abstract: Bio-tribology has achieved great prominence in the last few years as a new interdisciplinary field in which contributions from engineers, medical doctors, biologists, chemists, and physicists are co-ordinated. The main purpose of using a joint prosthesis is pain relief and restoring the joint function. To do this, a suitable material that has an infinite life is desirable. Material selection and component design are important factors in the performance and durability of total joint replacements but, unfortunately, wear of hip and knee bearings exists and is a significant clinical problem. There is a need not only for more wear-resistant materials but also for concomitant improvements in the design and manufacture of the implant and the operative techniques to minimize the occurrence of wear. Wear simulation is an essential pre-clinical method to predict the mid- and long-term clinical wear behaviour of prostheses, and one of the most important aspects of wear testing is simulation of actual wear conditions.

Keywords: bio-tribology, wear simulation, joints simulators, hip prostheses, knee prostheses, biological aspects.

6.1 Bio-tribology

The accepted worldwide definition of tribology is: 'The science and engineering of interacting surfaces in relative motion. It includes the study and application of the principles of friction, lubrication and wear.' The word 'tribology' derives from the Greek root ($\tau\rho\iota\beta$) of the verb ($\tau\rho\acute{\iota}\beta\omega$–tribo) and the suffix – logy (Dowson, 1998; Wikipedia, 2010c).

For centuries there was no word to describe the scientific concepts of friction, wear, and lubrication. The concept of tribology was enunciated in 1966 in a report of the UK Department of Education and Science (Bhushan, 1999). It encompasses the interdisciplinary science and technology of interacting surfaces in relative motion and associated subjects and practices. It includes parts of physics, chemistry, solid mechanics, fluid mechanics, heat transfer, materials science, lubricant rheology, reliability and performance. Although the name tribology is new, the constituent parts of tribology (wear, friction, and lubrication) are as old as history. During these interactions, forces are transmitted, mechanical energy is converted, and physical and chemical natures including surface topography of the interacting materials are altered (Bhushan, 1999).

Historically, Leonardo da Vinci (Dowson, 1998; Wikipedia, 2010c) was the first to enunciate two laws of friction. According to da Vinci, the frictional resistance was the same for two different objects of the same weight but making contact over different widths and lengths. He also observed that the force needed to overcome friction is doubled when the weight is doubled. The first reliable test on frictional wear was carried out by Charles Hatchett (1760–1820) using a simple reciprocating machine to evaluate wear on gold coins. He found that coins with grits between, compared to self-mated coins, wore at a faster rate (Eurotrib.org, 2010).

High friction is desirable between the foot and the floor for walking, whereas low friction is necessary for effortless flow of arterial blood cells. Surface contacts are likely to be unnoticed until they break down or become impaired following damage or disease (Neu et al., 2008). However, understanding the nature of these interactions and solving the technological problems associated with the interfacial phenomena constitute the essence of tribology, and understanding these principles is essential for the successful design of machine elements.

Usually, tribology is associated with the control of friction and wear in mechanical systems. However, these aspects are also a key factor in many biological functions. A wide range of examples can be considered, like hip and knee prosthetics, dental tissue and restorative materials, skin, hair and heart valves. Bio-tribology embraces all of these topics and has achieved great prominence in the last few years as a new interdisciplinary field in which contributions from engineers, medical doctors, biologists, chemists, and physicists are co-ordinated (Neu et al., 2008). This bioscience field has emerged from the classical field of tribology and is of paramount importance to the normal function of numerous tissues, including articular cartilage, blood vessels, heart, tendons, ligaments, and skin.

6.2 Basic concepts of anatomy and physiology of hip and knee joints

An understanding of the anatomy and the biomechanics of the joints is vital to replicate the wear pattern of artificial joints *in vitro*. Biomechanical principles (described in reference to the kinematics and kinetics of the joints) provide a valuable perspective to our understanding of the mechanism of artificial joint wear and thereafter the biological failure of prosthetic implants. Joint kinematics is the description of the angular or translational motion of the joint in response to applied forces; kinetics refers to the forces and moments acting on the joint during motion. To describe the kinematics of a joint three basic patterns of motion can be described: sliding, rolling, and spinning. Sliding (gliding) motion is defined as the pure translation of a moving segment against the surface of a fixed segment that has a constantly changing contact point. If the surface of the fixed segment is flat, the instantaneous centre of rotation is located at infinity. Otherwise, it is located at the centre of the curvature of the fixed surface. Spinning motion (rotation) is the exact opposite of sliding motion. In this case, the moving segment rotates, and the contact point on the fixed surface does not change. The instantaneous centre of rotation is located at the centre of the curvature of the spinning body that is undergoing pure rotation. Rolling motion occurs between moving and fixed segments where the contact points in each surface are constantly changing and the arc length of contact are equal on each segment. The instantaneous centre of rolling motion is located at the contact point. Most planar motion of anatomic joints can be described by using any two of these three basic descriptions.

6.2.1 Anatomy and biomechanics of the hip

The hip joint is one of the most stable joints in the body. The hip joint is a structure of four bones, forming a ball and socket joint between the pelvis and the femur (thigh), Fig. 6.1. The stability is provided by the rigid ball-and-socket configuration (enarthrodial), formed by the reception of the

Figure 6.1 Anatomy of the human hip joint

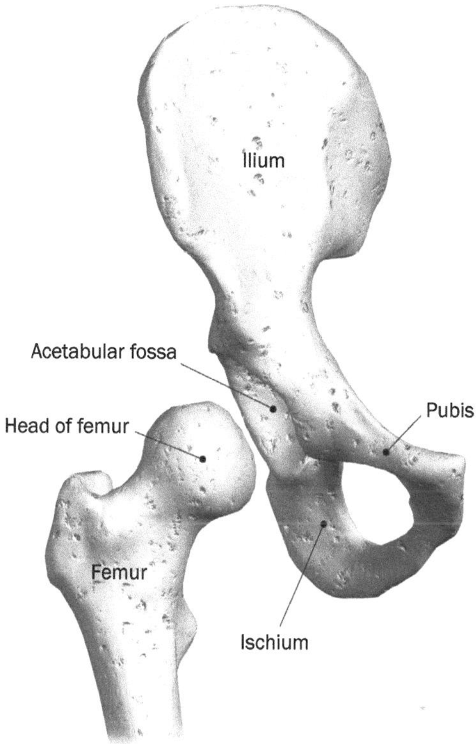

head of the femur into the cup-shaped cavity of the acetabulum. Although the movements of the hip are very extensive (the hip has three axes and three degrees of freedom), and consist of flexion, extension, adduction, circumduction, and rotation, the joint kinematics can be basically considered a pure spinning motion. In fact, in the hip joint the head of the femur is closely fitted to the acetabulum for a distance extending over nearly half a sphere, and is closely embraced by the acetabular labrum (a ring of cartilage that surrounds the acetabulum) that assures a seal effect to hold in place the joint. The maximum total range of motion of the hip joint is approximately 140° of flexion-extension, 75° of abduction-adduction, and 90° of rotation. The features of the hip joint derive from the two basic functions of the lower limb: support of the body weight and locomotion (Gray et al., 1974).

During the standing phase of gait, the entire articular surface of the acetabulum is involved in weight bearing and approximately 70–80% of the femoral head is in contact with the acetabulum. In the swing phase of gait, the dome of the acetabulum is no longer loaded, and only the anterior and posterior aspects of the femoral head are in contact. Contact pressures were found to be as high as 18 MPa in the postero-superior region of the acetabulum when rising from a chair (Hodge et al., 1989). The transition from incongruence during swing phase to congruence during load bearing has been shown to create high pressures in the hip, up to 330lb/in during gate.

The importance of the hip joint is not confined to the range of motion that it permits the upper leg, but also through the considerable muscular power and endurance that is delivered in concert with the motion. The hip has multiple muscle attachments (back, abdomen, hamstrings, quadriceps, abductors and adductors, and gluteal muscles). Most of the muscles of the hip are shorter and fatter than those of the leg, and allow rotation and help stabilize the joint. To provide a crude estimate of muscles and joint forces, the static loading of the hip joint has been frequently approximated with a simplified frontal lane analysis. In the two-dimensional static analysis of one legged-stance, the hip joint can be reduced to a simple lever arm system on which all forces acting parallel in the anatomic angles are ignored (Blount, 1956). In this model, the force exerted by the abductor muscle must produce a moment of equal magnitude, but in the opposite direction, to that produced by the effective body weight (BW) acting on the head of the femur. The relative ratio of length of the lever arm of the muscle to body weight is generally considered three to one, thus there is a mechanical advantage of three for the body weight force versus muscle force. For a person weighing 60 kg (about 600 N) the abductor muscle force would be three times the body weight minus the weight of the lower extremity below the hip joint (more or less 1/6 of the body weight), thus it would be three times 500 N or 1,500 N. To compute the joint reactive forces, it must be realized that a fulcrum force must act upward and be equal to the sum of the two forces acting downward if the system remains static and the forces are to be balanced. Accordingly, the total load on the fulcrum (the hip joint), would be approximately 500 N + 1,500 N = 2,000 N, which is just over three times body weight assuming the three to one lever arm ratio.

Important data have been obtained from instrumented joint prostheses (Davy et al., 1988; Bergmann et al., 2001). The average patient loads their hip joint with 238% BW (per cent of body weight) when walking at about four km/h and with slightly less when standing on one leg. When climbing upstairs, the joint contact force is 251% BW which is less than 260% BW when going downstairs. Inwards torsion of the implant (probably critical for the stem fixation) on average is 23% larger when going upstairs than during normal level walking.

6.2.2 Anatomy and biomechanics of the knee

Although the knee joint may look like a simple joint, it is one of the most complex. Four bones contribute to the joint. The femur (which is the bone of the thigh) attaches by ligaments and a capsule to the tibia (the larger bone of the leg). Next to the tibia is the fibula (the smaller bone of the leg), which runs parallel to the tibia. Finally, the patella is the bone in front of the knee (Fig. 6.2).

The knee joint must be regarded as consisting of three articulations in one: two between the femoral condyles and the corresponding tuberosity of the tibia (condyloid joints), and one between the patella and the femur. The bones of the knee are connected together by the ligaments. There are two cruciate ligaments located in the centre of the knee joint: the anterior cruciate ligament (ACL) prevents the femur from sliding backwards on the tibia (or the tibia sliding forwards on the femur), and the posterior cruciate ligament (PCL) which prevents the femur from sliding forward on the tibia (or the tibia from sliding backwards on the femur). Both ligaments stabilize the knee in a rotational fashion. The

Figure 6.2 Anatomy of the human knee joint

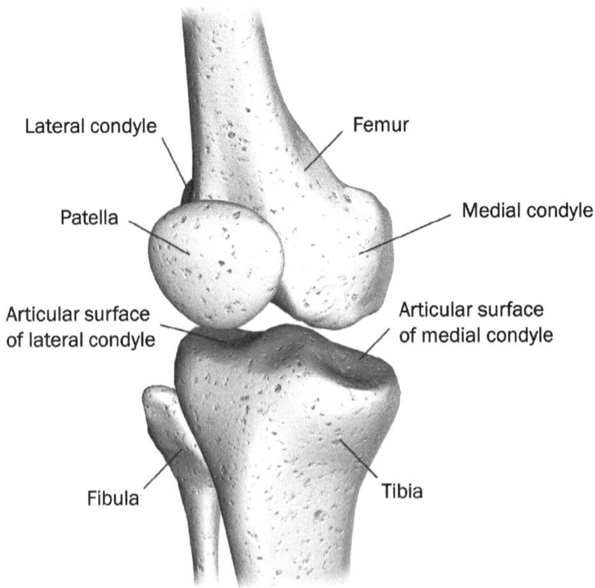

collateral ligaments (medial collateral ligament MCL, and lateral collateral ligament LCL) originate from the distal part of the femur and run distally to insert to the proximal part of the tibia (Fig. 6.3).

The primary function of collateral ligaments is to restrain the valgus and varus. The knee muscles which go across the knee joint are the quadriceps and the hamstrings. The quadriceps are on the front of the knee and attach on the proximal pole of the patella, the hamstrings are on the medial-posterior side of the knee and attach on the proximal medial part of the tibia.

The motion of the knee joint is polycentric and has six degrees of freedom; it is determined by the shape of the articulating surfaces of the tibia and the femur and the orientation of the four major ligaments of the knee joint. It consists essentially in the movement of flexion-extension, and, in certain positions, of slight rotation inward and outward.

Figure 6.3 **Components of human knee joint**

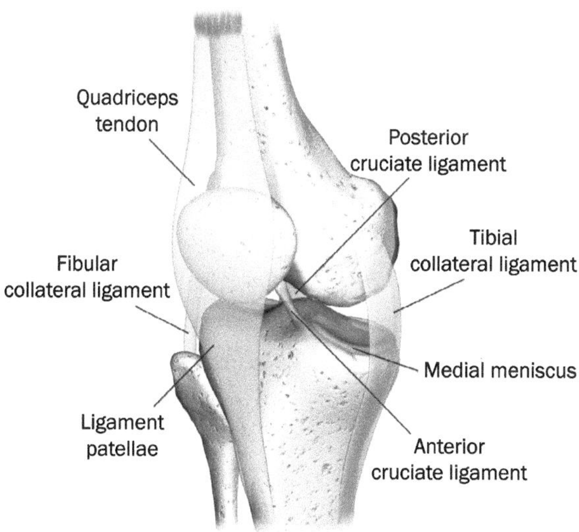

The movement of flexion-extension does not take place in a simple hinge-like manner, but is a complicated movement, consisting in a certain amount of sliding and rolling; so that the axis of motion is not a fixed one. Furthermore, during extension, the tibia externally rotates round a vertical axis drawn through the centre of the tibia. During flexion-extension, the patella moves on the distal part of the femur. In flexion only, the upper articular surface of the patella is in contact with the condyles of the femur; in the semi-flexed position the medial part of the patella is in contact with the femur; while in full extension, the patella is drawn up so that only the lower articular surface is in contact with the condyles of the femur. The patello-femoral joint has been described as having four mechanical functions: it increases the effective lever arm of the quadriceps; it provides functional stability under load; it allows the transmission of the quadriceps force to the tibia; and it provides a bony shield to the femoral troclea and condyles.

To further stabilize the joint during motion and to distribute uniformly the load bearing during loading, there are two semilunar fibrocartilage interposed between the femoral condyles and the tibia articular surfaces: the menisci. Experimental studies (Renström and Johnson, 1990; Ihn et al., 1993) have shown that in the absence of the menisci the load bearing area approximates $2\,cm^2$, and that it increases to $6\,cm^2$ on each condyle when the menisci are present. Besides, the menisci seemed to exert some stabilizing effect for both anterior–posterior and rotational displacements near full extension under a compressive load.

Knee stability results from a complex interaction among ligaments, muscles, menisci, the geometry of the articular surfaces, and the tibio-femoral reaction forces during weight-bearing activities.

6.3 Brief history of hip and knee prostheses

6.3.1 Hip history

Over the last three centuries, treatment of hip arthritis has evolved from rudimentary surgery to modern total hip arthroplasty (THA), which is considered one of the most successful surgical interventions ever developed (Gomez and Morcuende, 2005). Surgeons have been trying for well over a century to successfully treat this debilitating disease. Initial attempts to treat arthritic hips included arthrodesis (fusion), osteotomy, nerve division, and joint debridements. The goal of these early debridements was to remove arthritic spurs, calcium deposits, and irregular cartilage in an attempt to smooth the surfaces of the joint (Anonymous, 2010b). There was a great search for some material but surgeons and scientists were unable to find any which were biocompatible with the body, and yet strong enough to withstand the tremendous forces placed on the hip joint. Many attempts for hip arthroplasty were made with various materials from 1820 to 1940 using ivory, stainless steel, or moulding a 'piece of glass' into the shape of a hollow hemisphere, which could fit over the ball of the hip joint and provide a new smooth surface for movement (Barton, 1827; Rieker, 2003; Gomez and Morcuende, 2005; Anonymous, 2010b). While proving biocompatible, the glass could not withstand the stress of walking, and quickly failed.

A first significant improvement in this matter came in 1923 when a surgeon in Boston, Smith-Peterson, used a moulded glass cup to cover the reshaped head of the femur (Heybeli and Mumcu, 1999). Subsequently, other materials were tried until the manufacture of a cobalt-chromium alloy which was almost immediately applied to orthopaedics (Eftekhar and

Coventry, 1992). This new alloy was both very strong and resistant to corrosion, and has continued to be employed in various prostheses since that time.

However, the stage was set for Sir John Charnley to drive the evolution of a truly successful operation in orthopaedics, modern Total Hip Arthroplasty. In 1958, Charnley first reported his clinical experience with the replacement of a human joint using a steel femoral component and Teflon (Older, 2002; Gomez and Morcuende, 2005); unfortunately most of these prostheses failed. The first modern hip prosthesis was implanted in 1962 by Sir John (Rieker, 2003; Gomez and Morcuende, 2005), who developed the concept of 'low friction arthroplasty': a cemented stem with a 22.22 mm head in stainless steel combined with a cup made of polyethylene (UHMWPE). With the use of orthopaedic cement, metal-on-metal articulations (in the 1960s) and alumina-on-alumina articulations (in the 1970s) were also developed. Due to the better short-term results of low friction arthroplasties, these alternative bearings almost disappeared in the 1980s.

6.3.2 Knee history

A parallel line of development with the hip occurred with total knees. The first attempt at total knee arthroplasty was a prosthesis which was really a hinge fixed to the bones with stems into the medullary canals (the hollow marrow cavity) (Potter et al., 1972; Anonymous, 2010b; Wikipedia, 2010b). After a few short years, this prosthesis showed severe problems with loosening and infection and was abandoned. During this same period of time, some surgeons were trying to treat arthritis of the knee with a metal spacer, which was placed between the bones of the knee to eliminate the rubbing of irregular surfaces on each other. These implants, the McKeever in 1957 (McKeever, 1960) and the Macintosh in 1958 and 1964 (Macintosh, 1958), achieved some success but were not predictable, and many patients continued with significant symptoms.

Primitive replacements evolved from 1940 to 1965. The first, in the 1940s, involved a prosthesis that was hardly more than a hinge held in place by stems that extended into the hollow marrow cavities of the bones. Other attempts included metal spacers placed between the worn joints and moulds placed over the femoral halves of the knee bones. None were very successful. Then, in 1968, Frank Gunston, a Canadian orthopaedist, performed the first replacement operation using metal and plastic secured by surgical cement, a technique that has been perfected and is still the standard today (Anonymous, 2010b). In 1972, John Insall designed what has become the prototype for current total knee replacements (Anonymous, 2010b). This was a prosthesis made of three components, which would resurface all three surfaces of the knee – the femur, tibia and patella (kneecap). They were each fixed with bone cement and the results were outstanding. This was the first total knee complete with specific instrumentation to help with accurate bone cutting and implantation. Subsequently, the condylar knee was developed and the concept of replacing the tibiofemoral condylar surfaces with cemented fixation, along with preservation of the cruciate ligaments, was developed and refined (Ranawat, 2002). Condylar knee designs were improved to include modularity and non-cemented fixation, with use of universal instrumentation. However, significant advancements in the knowledge of knee mechanics and in the type and quality of the materials used (metals, polyethylene, and, more recently, ceramics) led to improved longevity (Ranawat et al., 1993; Deirmengian and Lonner, 2008; Lee and Goodman, 2008).

Current research with total knee replacement is directed at refining the design to improve patient function and the desire to achieve greater knee motion and strength motivates researchers to further enhance knee replacements so as to be equal to normal knees. Cementless fixation using prosthesis with a textured, porous surface into which bone can grow may provide biologic fixation, that is, the bone grows into the prosthesis and holds it in place. This may be more durable than cement used in the past. Cementless total knee arthroplasty is currently being used in patients, and the results look very promising.

6.4 Biomaterials used in hip and knee prostheses

The main purpose of using a joint prosthesis is pain relief and restoring the joint function. To do this, a suitable material that has an infinite life is desirable.

During the last 90 years, materials and devices have been developed to the point at which they can be used successfully to replace parts of living systems in the human body. These special materials, able to function in intimate contact with living tissue, with minimal adverse reaction or rejection by the body, and intended to interact with the biological system, are called biomaterials. Devices engineered from biomaterials and designed to perform specific functions in the body are generally referred to as biomedical devices or implants. Moreover, such materials have been biocompatible; the ability to perform with an appropriate host response in a specific application (Williams, 1986; Chow and Gonsalves, 1996). Design, material selection and biocompatibility remain the three critical issues in today's biomedical implants and devices. As advances have been made in the medical sciences, and with the advent of antibiotics, infectious diseases have become a much smaller health threat, but because average life expectancy has increased, degenerative diseases are a critical issue, particularly in the ageing population. More organs, joints, and other critical body parts will wear out and must be replaced if people are to maintain a good quality of life and biomaterials now play a major role in replacing or improving the function of every major body system (skeletal, circulatory, nervous, etc.). Some common implants include orthopaedic devices such as total knee and hip joint replacements, spinal implants and bone fixators, cardiac implants such as artificial heart valves and pacemakers, soft tissue implants such as breast implants and injectable collagen for soft tissue augmentation, and dental implants to replace teeth/root systems and bony tissue in the oral cavity.

Material choices must take into account biocompatibility with surrounding tissues, the environment and corrosion issues, friction and wear of the articulating surfaces, and implant fixation either through osseo integration (the degree to which bone will grow next to or integrate into the implant) or bone cement. In fact, the orthopaedic implant community agrees that one of the major problems plaguing these devices is purely materials-related: wear of the polymer cup in total joint replacements (Brinker and Sherrer, 1990). The average life of a total joint replacement is 8–12 years (Matijevic, 1985), even less in more active or younger patients. There are growing numbers of younger and more active patients who require total hip and knee replacement, for example, as a result of skiing or motorcycle accidents. Their increased activity plus longer usage is expected to result in a higher incidence of eventual failure of conventional hip and knee replacements. Because it is necessary to remove some bone surrounding the implant, generally only one revision surgery is possible, thus limiting current orthopaedic implant technology to older, less active individuals.

The materials properties required for articulating surfaces in combination with load bearing applications are:

- High long-term mechanical strength, i.e. tensile and compressive strength combined with high fracture toughness and appropriate creep and fatigue resistance.
- Wear resistance, based on hardness and low roughness.
- No risk of failure *in vivo*.
- Biocompatibility with surrounding tissues.

Various materials have been proposed for balls, cups and stems. Different configurations for the articulating surfaces have been tested: metal-high density polyethylene, metal–metal, ceramic–polyethylene and ceramic–ceramic. Table 6.1 shows the different solutions studied up to now and the limiting factor affecting the lifetime. Questions remain, however,

Table 6.1 Bearing system proposed and their problems

Couple	Problem
Metal–polyethylene	Wear and fatigue-induced delamination of the polymeric component. Small submicron debris is believed to be responsible for adverse tissue reaction and subsequent osteolysis and implant loosening
Metal-on-metal	Significant amount of chromium, nickel, and cobalt is released in the body as a consequence of metal wear
Alumina–polyethylene	Fracture rates of up to 1.6% due to brittleness of alumina
Alumina–alumina	Higher fracture rates than in the case of alumina–polyethylene due to brittleness of alumina
Zirconia–polyethylene and Zirconia–zirconia	Hydrothermal degradation of zirconia

concerning which prosthetic designs and materials are most effective for specific groups of patients and which surgical techniques and rehabilitation approaches yield the best long-term outcomes. Issues also exist regarding the best indications and approaches for revision surgery. Below a description of the biomaterials currently used in prosthetic implants.

6.4.1 Polyethylene (UHMWPE)

Polyethylene (UHMWPE) is currently adopted in 1.4 million patients around the world every year for use in the hip, knee, upper extremities and spine. It has been used in hip replacements for over forty years (Bellare et al., 2005; Devine, 2006) and Fig. 6.4 shows some components.

Although the choice of this material is very common, the life of artificial joints is limited to approximately ten years, after which they decline markedly. Recently, the orthopaedic industry has developed new processing techniques, such as radiation crosslinking, which are expected to dramatically reduce wear and improve the longevity of hip implants beyond ten years (Kurtz, 2004).

The generic formula of polyethylene is $-(C_2H_4)_n-$; for UHMWPE the molecular chain can consist of as many as

Figure 6.4 Polyethylene components used in hip and knee orthopaedics implants

200,000 ethylene repeat units. In general terms, after the polymerization from ethylene gas, UHMWPE in the form of resin powder needs to be consolidated in the form of sheet, rod or in the form of near net implant. Finally, in most instances, the UHMWPE implant must be machined into its final shape (Kurtz, 2004). UHMWPE is a semi-crystalline, two-phase viscoplastic solid composed of an amorphous matrix responsible for the resistance to mechanical deformations, embedding crystalline domains (Renò and Cannas, 2006). The crystalline phase consists of folded rows of carbon atoms packed into lamellae. The surrounding amorphous phase consists of randomly oriented and entangled polymer chains traversed by tie molecules, which interconnect lamellae that provide resistance to mechanical deformation. UHMWPE can evolve over time in response to its mechanical, chemical and thermal history. It has long been known that processing can substantially influence the morphology, and hence the mechanical behaviour of UHMWPE (Edidin et al., 1999; Reggiani et al., 2006).

Crosslinking has been extensively introduced to reduce the wear of UHMWPE. Following the identification of the acceleration of wear due to oxidative degradation of UHMWPE which had been gamma sterilized in air, stabilized UHMWPE (which is irradiated in an inert atmosphere) and intentionally crosslinked UHMWPE have been developed and introduced clinically (Wikipedia, 2007; Jacobson, 2008). Whereas cross linking is essential for reducing abrasive and adhesive wear, elimination of the free radicals is an important step in reducing the long-term oxidative degradation and embitterment of the UHMWPE, but UHMWPE introduces other modes of damage such as delamination and fatigue, which could adversely affect the device performance (Shen and McKellop, 2002; Anonymous, 2008).

6.4.2 Metal

The metal-on-metal (MOM) concept was introduced in the 1960s as the so-called McKee–Farrar prosthesis. It was introduced to solve the critical problems of polyethylene wear (Rose et al., 1980; Willmann, 1998) and Fig. 6.5 shows some components.

MOM articulation is typically associated with the cobalt-chromium-molybdenum alloy (CoCrMo according to ISO 5832-4). Typically, these alloys are divided into two categories: high carbon alloys, where the C content is above 0.20%; and low-carbon alloys (carbon content < 0.08%) (standard ASTM F75-07).

CoCrMo implants can be manufactured using two different techniques: casting and forging. The grain size of the forged alloy is typically less than 10 μm, whereas the grain size of the cast material ranges from 30 to 1000 μm (Asphahani, 1987; Wang et al., 1999). The carbides are also smaller in forged components than in the cast components. The effects of these variations in material in the manufacturing process

Figure 6.5 Metallic components used in hip and knee orthopaedics implants

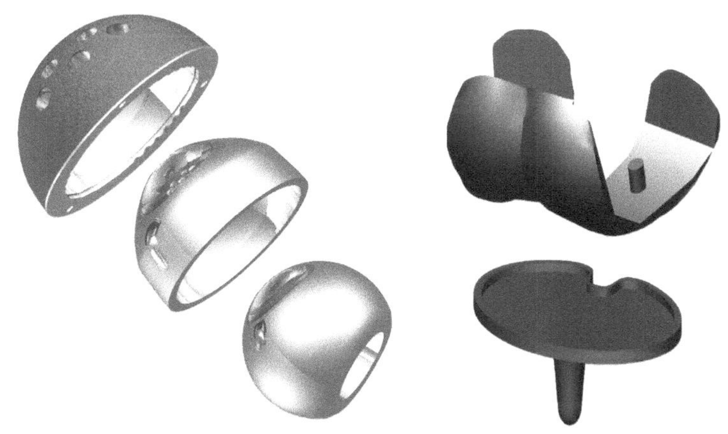

have effects on the wear rate, on the production of particles debris, and on the micro-structure of the CoCrMo alloy (Tipper et al., 2005). A high carbon content improves wear resistance in cast implants because it increases material hardness (Mcminn, 2003) while a low carbon content is preferable with forged components (Chan et al., 1996, 1999). However, even if low carbon metal-on-metal bearings have demonstrated a good *in vitro* tribological behaviour, the clinical behaviour seems to be considerably worst than high carbon metal-on-metal bearings.

Very large grains, as can occur in cast cobalt-chromium alloy, decrease yield strength which can lead to catastrophic failure in load-bearing implants; in addition, the large carbides typical of these alloys can also be removed from the surface, creating third body wear conditions (Asphahani, 1987; Mancha et al., 1996). Corrosion of metals is the most obvious form of degradation (Morais et al., 1998, 1999), and this is defined as the unwanted chemical reaction of a metal with its environment, resulting in its continued degradation to oxides, hydroxides or other compounds (Black, 1996). Moreover microstructure, morphology, carbide fraction, and diameter size may all influence wear rate and corrosion properties of base alloys (Marti, 2000; Buscher et al., 2005; Saldivar-Garcia and Lopez, 2005). The goal of MOM-bearing combinations is to reduce wear to less than a clinically relevant level, that is a level that does not induce osteolysis or another outcome that necessitates revision surgery.

A number of clinical studies have shown that some of these MOM prostheses can last twenty or even up to thirty years, while others can fail relatively early (Wimmer et al., 2003). In 1988, the Metasul second-generation MOM hip prosthesis was introduced into clinical practice, comprised of a cobalt chrome alloy femoral head articulating on a cobalt-chromium alloy acetabular cup. Over 200,000 Metasul combinations have been implanted to date (Tipper et al., 2005) and short-term clinical performance has been encouraging, with low wear rates and few prostheses requiring revision.

Metals react with the oxygen-rich biological environment and form a thin protective oxidative coating generally 2–5 nm thick (Wikipedia, 2010a) that prevents corrosion. The oxidative film forms instantly once exposed to the *in vivo* conditions, but is not permanently fixated on the metals. The coatings are capable of being scratched or rubbed off when undergoing surface contact and once the coating is dissipated, the implant metals are susceptible to releasing metal ions and particulates. The presence of the particulate and ions creates third body wear that dramatically increases wear rates due to the substantial increase in roughness. This detrimental cycle applied to the coating, to the metal ions released, and to the reformation of new coatings is referred to as oxidative wear (Davidson, 1993). While metallic implants show ion release of various types, ceramics show no signs thereof. The highest ion release was shown by stainless steel (SS 316). *In vitro* testing and examination of retrieved implants and synovial fluid suggest that modern metal-on-metal bearings produce minimal early wear and a lower incidence of periprosthetic osteolysis than metal-on-polyethylene implants (Savarino et al., 2006).

6.4.3 Ceramic

Ceramic materials for orthopaedic applications are a class of new engineering materials for wear resistant applications under severe environments. In the field of materials science, the term 'ceramic' includes all non-metallic, inorganic materials; this word is derived from the Greek, *κεραμοσ* and includes glass products, cement and plasters, some abrasive and cutting tool materials, various electrical insulation materials, porcelain and other refractory coatings for metals, etc. (Ravaglioli and Krajewski, 1992; Hsu, 1996). Ceramic materials for hip joints were introduced more than thirty years ago to solve the critical problems of polyethylene wear (Cuckler et al., 1995; Toni et al., 1995) and the first ceramic-on-ceramic coupling was implanted in France by Dr. Boutin (Boutin, 1972).

Interest in ceramics for biomedical applications has increased over the last thirty years and the properties to use these materials for an orthopaedic application follows the international guideline ISO 6474-10. The ceramics that are used in implantation and clinical purposes include aluminium oxide (alumina), partially stabilized zirconia (PSZ) (both yttria tetragonal zirconia polycrystal [Y-TZP] and magnesia partially stabilized zirconia [Mg-PSZ]), bioglass®, glass-ceramics, calcium phosphates (hydroxyapatite and β-tricalcium phosphate) and crystalline or glassy forms of carbon and its compounds. Figure 6.6 shows some components.

The theoretical advantages of alumina ceramic are related to the excellent biocompatibility, exceptional tribological

Figure 6.6 Ceramic components used in hip and knee orthopaedics implants

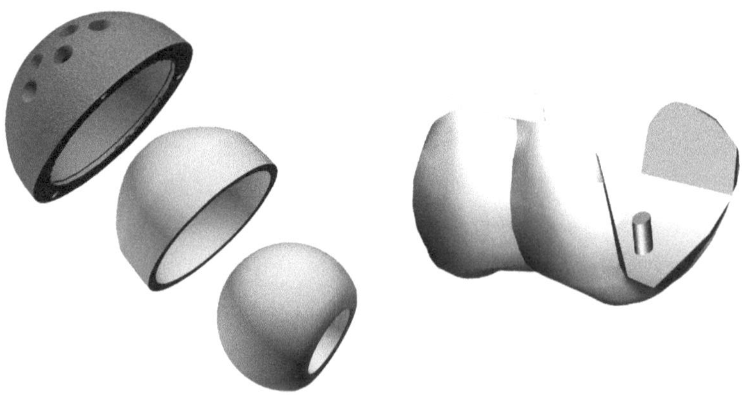

properties due to high scratch resistance and wettability of the material, good properties of mechanical and wear resistance, good properties at high temperatures where metals cannot be used, and chemical stability (McKellop et al., 1992; Toni et al., 1995; Hsu, 1996; Willmann, 1998). Zirconia ceramics have been introduced into orthopaedics as an alternative to alumina (Derbyshire et al., 1994). They have several advantages over other ceramic materials due to the transformation toughening mechanisms operating in their microstructure that give their components very interesting mechanical properties (Piconi and Maccauro, 1999). Alumina is chemically more stable but it is mechanically weaker than zirconia; however, the degradation of zirconia, in which the phase transformation is accelerated in an aqueous environment, is a limitation to this material in bioceramic use (Piconi et al., 1998; Piconi and Maccauro, 1999).

Current commercially available components are generally composed of a single medical grade ceramic, therefore scientists developed the idea of combining alumina and zirconia in different percentages. A fraction of zirconia added to alumina reduces the incidence of fracture and results in a composite material of increased toughness (Cherif et al., 1996; Affatato et al., 1999, 2001; Piconi and Maccauro, 1999). This may be a particular solution to obtain a new material with improved mechanical and tribological properties than the pure ceramics components, commonly perceived as rigid and subject to brittle fracture. Biolox Delta is the brand name of this third generation of ceramics in which the resistance and toughness of alumina is combined with the resistance and toughness of zirconia in order to create a composite material with good biocompatibility, chemical and hydrothermal stability, high resistance to wear and good mechanical features (toughness and strength) (Dalla Pria and Burger, 2003; Benazzo et al., 2007).

6.5 Wear of biomaterials

Material selection and component design are important factors in the performance and durability of total joint replacements but, unfortunately, wear of hip and knee bearings exist and is a significant clinical problem. The wear of the implant products can cause adverse tissue reaction that may lead to massive bone loss around the implant and consequently loosening of the fixation. A schematization of this phenomenon is better emphasized in Fig. 6.7.

The tribological interactions of a solid surface's exposed face with interfacing materials and environment may result in loss of material from the surface, known as 'wear'. Wear is the erosion of material from a solid surface by the action of another surface. It is related to surface interactions and more specifically to the removal of material from a surface as a result of mechanical action (Rabinowicz, 1995; Czichos, 1997). Wear can be minimized by modifying the surface properties of solids by one or more surface finishing or by use of lubricants (for frictional or adhesive wear), but any material can wear by a lot of movement influenced by

Figure 6.7 Schematization of osteolysis phenomenon due to wear and particles debris

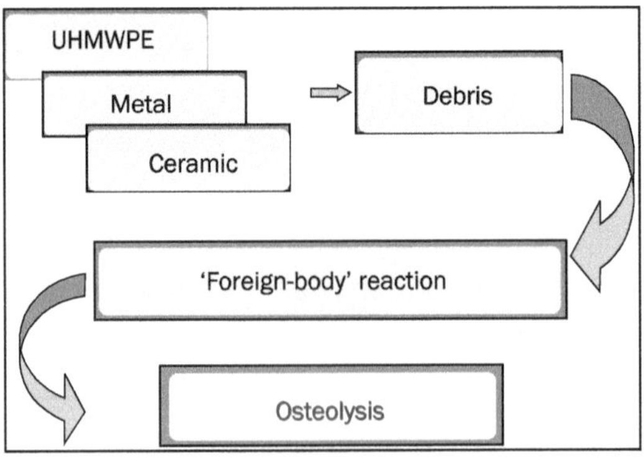

factors such as ambient conditions, temperature, loading, counterface, etc. (Bayer, 1997). Many physical wear mechanisms exist and it must be emphasized that there is more than one distinct mechanism of wear, with sensitivity to parameters such as load, speed, etc., being different for different wear situations (Bayer, 1997). Major types of wear include abrasion, adhesion (friction), erosion, fretting, and corrosion.

- *Abrasion:* Abrasive wear occurs when a hard rough surface slides across a softer surface. In other words it is defined as the loss of material due to hard particles or hard protuberances that are forced against and move along a solid surface (Rabinowicz, 1995 and standard ASTM-G40, 2009). Abrasive wear is commonly classified according to the type of contact and the contact environment. The two modes of abrasive wear are known as two-body and three-body abrasive wear. Two-body wear occurs when the grits, or hard particles, are rigidly mounted or adhere to a surface, and material is removed from the surface. The common analogy is that of material being removed with sandpaper. Three-body wear occurs when the particles are not constrained, and are free to roll and slide down a surface. The contact environment determines whether the wear is classified as open or closed. An open contact environment occurs when the surfaces are sufficiently displaced to be independent of one another (Wikipedia, 2010d).
- *Adhesion:* The tendency of certain dissimilar molecules to cling together due to attractive forces. Adhesive wear occurs when two bodies slide over each other, or are pressed into one another, which promote material transfer between the two surfaces. However, material transfer is always present when two surfaces are aligned

against each other for a certain amount of time (Rabinowicz, 1995; Wikipedia, 2010d).

- *Erosion:* A gravity driven process that moves solids (sediment, soil, rock and other particles) in the natural environment or their source and deposits them elsewhere (Wikipedia, 2009). It usually occurs due to transport by wind, water, or ice; by down-slope creep of soil and other material under the force of gravity; or by living organisms, such as burrowing animals, in the case of bio-erosion. The impacting particles gradually remove material from the surface through repeated deformations and cutting actions.
- *Fretting wear:* The repeated cyclical rubbing between two surfaces, which is known as fretting, over a period of time which will remove material from one or both surfaces in contact (Wikipedia, 2008). It occurs typically in bearings, although most bearings have their surfaces hardened to resist the problem. Another problem occurs when cracks in either surface are created, known as fretting fatigue. It is the more serious of the two phenomena because it can lead to catastrophic failure of the bearing.
- *Corrosion:* The disintegration of an engineered material into its constituent atoms due to chemical reactions with its surroundings (Anonymous, 2010a). In the most common use of the word, this means electrochemical oxidation of metals in reaction with an oxidant such as oxygen. Formation of an oxide of iron due to oxidation of the iron atoms in solid solution is a well-known example of electrochemical corrosion, commonly known as rusting, and this type of damage typically produces oxide(s) and/or salt(s) of the original metal. Corrosion can also refer to other materials than metals, such as ceramics or polymers, although in this context the term

degradation is more common. In other words, corrosion is the wearing away of metals due to a chemical reaction.

6.6 Wear evaluation

Wear is not a basic material property, but a system response of materials. The America Society for Testing and Materials (ASTM) define in a standard guideline (ASTM G40-10) that 'wear is defined as damage to a solid surface, generally involving progressive loss of material . . .'. This statement means that the surfaces of any material can alter and damage because of an alteration to them. A single general-purpose wear test that establishes a unique wear parameter or rating of a material does not exist, so to assess the amount of wear, several approaches have been used to determine its value and to prescribe appropriate solution. Scientists have used wear testing in order to rank wear resistance of materials, and complex methods of wear testing were developed in order to determine wear parameters that can project performance and to establish the influence of various factors on these parameters (Bezing, 1973; Bayer, 1997).

Wear simulation is an essential pre-clinical method to predict the mid- and long-term clinical wear behaviour of prosthesis, and one of the most important aspects of wear testing is simulation of actual wear conditions. Generally, two categories of laboratory tests are conducted: wear-screening devices and wear joint devices.

6.6.1 *Wear screening devices*

These categories of tests, also called quick-tests, provide information exclusively on the intrinsic features of the materials studied, without reproducing either the features of the shape of the implant, or the environment with which it will have to interact, and they are short. Figure 6.8 shows a representation of some quick-tests.

They are quick, do not accurately represent the specimen geometry of the biomaterials used that can influence the lubrication or the contact stress, and do not reproduce accurately the wear mechanism operating *in vivo*. They are based on physical mechanisms of material removal or displacement and operational mechanisms such as mechanical action: rolling wear, sliding wear, material interaction, etc. They are less useful in predicting wear rates in the implanted joint, unless care is taken to simulate the loading cycle, all aspects of the motion between head and cup, body environment and the effective mode of lubrication (Dowson, 2001 and standards ASTM-F732-00, 2006 and ASTM-G133, 1995).

Figure 6.8 **Standard wear screening devices used in order to give information exclusively on the intrinsic features of the materials studied**

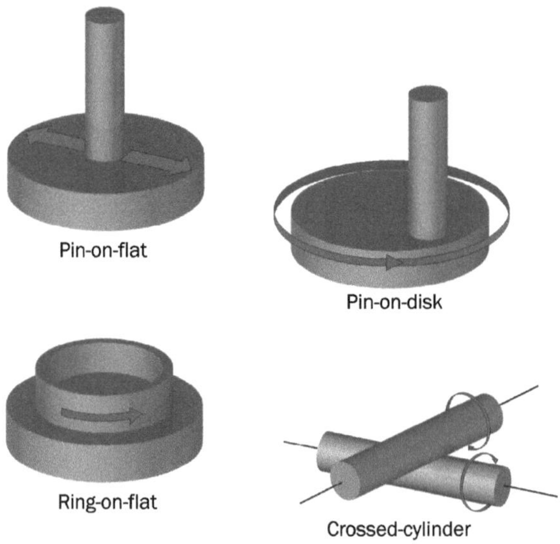

These devices have some disadvantages, such as the inability to extrapolate the results obtained and to predict wear of the specimens tested, and the wear behaviour may fail to predict *in vivo* results. In these machines, the wear mechanism may also be influenced by the great heat, generated by friction, which is transferred to the surfaces. In metal against polyethylene simulations, Davidson and co-workers (Davidson and Schwartz, 1987; Davidson et al., 1988) found that the temperature at the interface increased by about 10°C.

For orthopaedic applications, international guidelines (ASTM F & G) give some recommendations in order to conduct a wear test, but some of the most important quick-tests will be briefly described here.

- *Pin-on-disk:* This test consists of a laboratory procedure for determining the wear of materials during sliding using a pin-on-disk apparatus. Materials are tested in pairs under nominally non-abrasive conditions. This practice follows the guidelines contained in ASTM G99-05.
- *Pin-on-flat:* This test consists of a laboratory method for evaluating the friction and wear properties of combinations of materials (that are being considered for use as the bearing surfaces of human total joint replacement prostheses). This practice follows the guidelines contained in ASTM F732-06. It is used to rank the materials with regard to friction levels and polymer wear rate under simulated physiological conditions.
- *Block-on-ring:* This test consists of a laboratory procedure for determining the resistance of materials to sliding wear. The test utilizes a block-on-ring friction and wear testing machine to rank pairs of materials according to their sliding wear characteristics under various conditions. This practice follows the guidelines contained in ASTM G77-05.

- *Crossed-cylinder:* This test consists of a laboratory test for ranking metallic couples in their resistance to sliding wear using the crossed-cylinder apparatus. This practice follows the guidelines contained in ASTM G83-96.

6.6.2 Wear joint simulators

These machines represent a more complex dynamic situation of analysis than those above and typical hip joint and knee joint wear simulators are shown in Fig. 6.9 and Fig. 6.10. In these simulators, real prostheses are tested in an environment that simulates physiological conditions in order to predict some aspect of clinical performance of the materials tested *in vivo* wear patterns (Bragdon et al., 1996; Barbour et al., 2000; Goldsmith et al., 2000; Clarke et al., 2001; Saikko, 2005; Affatato et al., 2006).

The development of improved materials to extend the lifetime of orthopaedic implants, such as knees and hips, up

Figure 6.9 Schematic view of a typical hip joint wear simulator

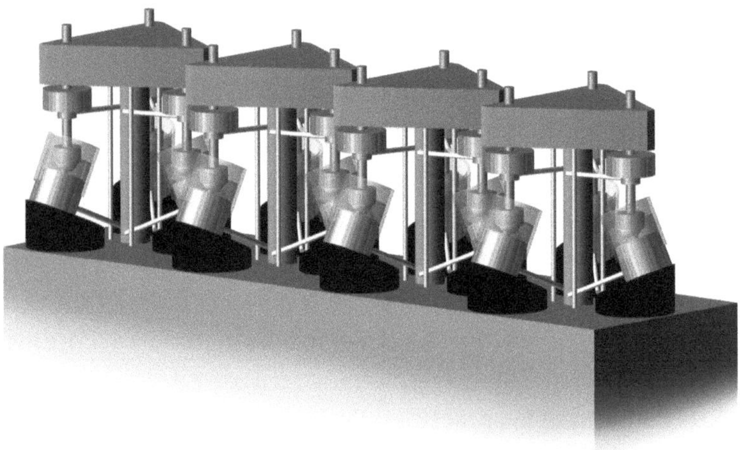

Figure 6.10 Schematic view of a typical knee joint wear simulator

to a minimum of thirty years is a critical social objective. Considering that all new materials have to be tested before clinical trials, joint simulators play an important role in this pre-clinical validation. These machines can also be used as research tools allowing experiments to be conducted in a relatively controlled environment where variables such as surface roughness and scratching can vary and the effects measured. To replicate/simulate particularly extreme conditions, a wear joint simulator could be of great help. In particular, simulator wear tests can be used to conduct accelerated protocols so that it is possible to establish the limits of performance for the material in short time frames.

Hip and knee wear simulators are widely used and have been successful in evaluating the wear properties of total hip/knee arthroplasty articulating surfaces. These test results have been shown to correlate well to clinical experience. Since this type of test machine is widely used, it is important that parameters can be standardized to make results more uniform and comparable over all laboratories. The simulators currently in use differ from each other in many parameters: number of stations, loading (physiological or simplified), degrees of freedom, anatomical or inverted configuration, and temperature-controlled test fluid baths for each hip joint assembly. Finally, yet important, is the simulator that allows independence among the stations (for example if one specimen needs to be removed for examination or replacement, the testing could be disabled while the remaining stations keep running).

For orthopaedic applications, international guidelines (ISO 14242 for hip and ISO 14243 for knee) give some recommendations in order to conduct a wear test. In particular, these international guidelines specify the relative angular movement between articulating components, the pattern of the applied force, speed and duration of testing, sample configuration and test environment to be used for the wear testing of total hip/knee joint prostheses. These wear tests need a complex set up because the test specimens have to be placed onto the simulator in a particular configuration. The test apparatus transmits a specified time-varying force between the components, together with specified relative angular displacements, and the test takes place in a controlled environment simulating physiological conditions. Normally the fluid test medium is filtered through a 2-μm filter, and has a protein mass concentration of not less than 17g/l as recommended by the aforementioned guidelines. To minimize microbial contamination, the fluid test medium should be stored frozen until required for testing. An anti-microbial reagent (such as sodium azide) may be added, but such reagents can be potentially hazardous.

The material loss (the wear) from components of the prosthetic joint, due to combined movement and loading, is evaluated using different methodologies, but the gravimetric way is considered the gold standard. As recommended by the aforementioned international guidelines, the test specimen is soaked in a lubricant (usually Bovine calf serum) and is repeatedly removed from the lubricant, cleaned, dried and weighed until a steady rate of fluid sorption is established. After recording the mass of the specimens and mounting them in the testing machine, the wear test is conducted in accordance with ISO 14242-1/2 (for hip specimens) and ISO 14243-1/2 (for knee specimens). The test specimen is assessed subsequently for wear by testing for loss in mass in a hip/knee simulator. A loaded, non-articulating control specimen is intended to allow for fluid sorption and undergoes the same procedure for reference purposes.

On each occasion when the test specimen and control specimen are removed from the wear-testing machine, clean and dry the components and calculate the gravimetric wear as follows:

$$Wn = Wan + Sn \qquad [6.1]$$

where Wn is the net mass loss after n cycles of loading, Wan is the average uncorrected mass loss, and Sn is the average increase in mass of the control specimen over the same period.

6.7 Biological effects of wear

Willert and Semlitsch in the 1970s (Willert et al., 1974; Willert and Semlitsch, 1977) proposed that aseptic loosening of joint prostheses resulted from the abundant wear debris seen microscopically around and within reactive blood cells (macrophages), which comprised the majority of the periprosthetic fibrous tissue membrane. They suggested that wear debris was biologically active and induced a macrophage response to the surrounding tissues. Since than, numerous investigators have shown that the cellular activity in the membrane is capable of producing a variety of enzymes, prostaglandins, and cytokines, which are capable of stimulating osteoclastic bone resorption and fibrous tissue formation (Goldring et al., 1983; Kim et al., 1993; Shanbhag et al., 1995; Tuan et al., 2008).

The factors that contribute to osteolysis (bone resorption around or at the implant–bone interface) are related to the number, size, shape, rate of generation, time of exposure, and antigenic properties of the wear debris particles. The macrophage is the predominant cell type with respect to biomaterial particles in inciting periprosthetic inflammatory bone loss. Other cells recognized as taking part in the process of osteolysis are fibroblasts, osteoblasts, osteoprogenitor cells (adult mesenchymal stem cells), synovial cells, and osteoclasts (Tuan et al., 2008). The activation process by which wear debris activates macrophages and other cells within the interfacial membrane is similar. During the life of an implant, and especially in the case of poor primary stability with a constant accumulation of wear particles over time, marrow cells are exposed to high concentrations of debris and also when implant fixation is at the beginning relatively weak. As a result of the prolonged exposure to particles, the normal osteogenic differentiation process may disrupt, diminishing the population of functional osteoblasts, thus compromising osseointegration at the bone–implant interface. *In vitro* evidence suggests that wear debris can alter osteoblast and osteoprogenitor cell function and increase osteoclast activity, resulting in decreased bone matrix production and osteolysis. Tuan et al. (2008) suggest a possible genetic predisposition to osteolysis on the basis of the clinical variation seen in the osteolytic response to implant wear. The heritable component should depend on multiple minor DNA sequence variations occurring with a stable frequency within the population, leading to subtle changes in gene function, giving rise to altered susceptibility of severity for osteolysis.

Clinically the bone loss appears as a linear, diffuse, dissecting phenomenon compromising the bone–implant interface, resulting in a generalized enlargement of the bone canal and endosteal bone lysis; alternatively, the bone loss can appear to be focal, manifesting as an endosteal lytic lesion. The end result is typically a loose prosthesis requiring revision surgery, or possibly a periprosthetic fracture.

6.8 Acknowledgements

The authors would like to thank Luigi Lena for the illustrations (Laboratorio di Tecnologia Medica, Rizzoli Orthopaedic Institute, Bologna) and Stefania Affatato for her help with the English language.

6.9 References

Affatato, S., Goldoni, M., Testoni, M. and Toni, A. (2001) 'Mixed oxides prosthetic ceramic ball heads. Part 3: Effect of the ZrO_2 fraction on the wear of ceramic on ceramic hip joint prostheses. A long-term in vitro study', *Biomaterials* 22: 717–23.

Affatato, S., Leardini, W., Jedenmalm, A., Ruggeri, O. and Toni, A. (2006) 'Larger diameter bearings reduce wear in metal-on-metal hip implants', *Clin Orthop Relat Res*, Epub ahead of print.

Affatato, S., Testoni, M., Cacciari, G. L. and Toni, A. (1999) 'Mixed-oxides prosthetic ceramic ball heads. Part 2: Effect of the ZrO_2 fraction on the wear of ceramic on ceramic joints', *Biomaterials* 20: 1925–9.

Anonymous (2008) 'Electron-Beam Processing of Plastics: An Alternative to Chemical Additives'. Available from: http://www.ebeamservices.com/ebeam_spe_antec.htm (7 February 2008).

Anonymous (2010a) 'Corrosion'. Available from: http://www.scribd.com/doc/24010913/The-Disintegration-of-an-Engineered-Material-Into-Its-Constituent-Atoms (5 March 2010).

Anonymous (2010b) 'History of Total Joint Replacement'. Available from: http://www.utahhipandknee.com/history.htm (1 March 2010).

Asphahani, A. I. (1987) 'Corrosion of cobalt base alloys', in Davis, J. R. (ed.), *Corrosion*. The materials information society, U.S.A.

Barbour, P. S., Stone, M. H. and Fisher, J. (2000) 'A hip joint simulator study using new and physiologically scratched femoral heads with ultra-high molecular weight polyethylene acetabular cups', *Proc Inst Mech Eng [H]* **214**: 569–76.

Barton, J. R. (1827) 'On the treatment of anchylosis, by the formation of artificial joints', *North Amer Med and Surg Journal* 3: 40.

Bayer, R. G. (1997) 'Wear testing', in International, A. (ed.), *Friction and Wear Testing*. ASM.

Bellare, A., Bistolfi, A., Simis, K. and Pruitt, L. (2005) UHMWPE processing: techniques and problems. *UHMWPE for Arthroplasty: Degradation, Stabilisation, and Crosslinking*. Turin, Italy.

Benazzo, F., Macchi, F., Rossi, S. and Dalla Pria, P. (2007) 'Ceramic total knee arthroplasty – an update', *Orthopaedic Surgery* 59–62.

Bergmann, G., Deuretzbacher, G., Heller, M., Graichen, F., Rohlmann, A. et al. (2001) 'Hip contact forces and gait patterns from routine activities'. *J Biomech* **34**: 859–71.

Bezing, R. (1973) *Friction and Wear Devices.*

Bhushan, B. (1999) 'Definition and history of tribology', in I. (ed.) *Principles and Applications of Tribology*, New York, John Wiley & Sons.

Black, J. (1996) 'Metal on metal bearings. A practical alternative to metal on polyethylene total joints?', *Clin Orthop* **329**: 244–55.

Blount, W. (1956) 'Don't throw away the cane', *J Bone Joint Surg Am* **38A**: 695–708.

Boutin, P. (1972) 'Ceramic/ceramic in total hip arthroplasty', *Rev Chir Orthop* **58**: 229–46.

Bragdon, C. R., O'Connor, D. O., Lowenstein, J. D., Jasty, M. and Syniuta, W. D. (1996) 'The importance of multidirectional motion on the wear of polyethylene', *Proc Inst Mech Eng [H]* **210**: 157–65.

Brinker, C. J. and Sherrer, G. W. (1990) *Sol-Gel Science*, New York, Academic Press.

Buscher, R., Tager, G., Dudzinski, W., Gleising, B., Wimmer, M. A. and Fischer, A. (2005) 'Subsurface microstructure of metal-on-metal hip joints and its relationship to wear particle generation', *J Biomed Mater Res B Appl Biomater* **72**: 206–14.

Chan, F. W., Bobyn, J. D., Medley, J. B., Krygier, J. J. and Tanzer, M. (1999) 'Wear and lubrication of metal-on-metal hip implants', *Clin Orthop & Rel Res* **369**: 10–24.

Chan, F. W., Bobyn, J. D., Medley, J. B., Krygier, J. J., Yue, S. and Tanzer, M. (1996) 'Engineering issues and wear performance of metal on metal hip implants', *Clin Orthop Rel Res* **333**: 96–107.

Cherif, K., Gueroult, B. and Rigaud, M. (1996) 'Wear behaviour of alumina toughened zirconia materials', *Wear* **199**: 113–21.

Chow, G. M. and Gonsalves, K. E. (1996) 'Particle synthesis by chemical routes', in Edelstein, A. S. and Cammarata, R. C. (eds.) *Nanomaterials: Synthesis*. Bristol and Philadelphia, Institute of Physics Publishing.

Clarke, I. C., Chan, F. W., Essner, A., Good, V., Kaddick, C. et al. (2001) 'Multi-laboratory simulator studies on effects of serum proteins on PTFE cup wear', *Wear* **250**: 188–98.

Cuckler, J. M., Bearcroft, J. and Asgian, C. M. (1995) 'Femoral head technologies to reduce polyethylene wear in total hip arthroplasty', *Clin Orthop Relat Res* **317**: 57–63.

Czichos, H. (1997) 'Basic tribological parameters', in International, A. (ed.) *Friction and Wear Testing*. ASM.

Dalla Pria, P. and Burger, W. (2003) 'Una nuova ceramica per l'ortopedia: BIOLOX® Delta', *Rivista di Patologia Dell'Apparato Locomotore* **2**: 1–6.

Davidson, J. (1993) 'Characteristics of metal and ceramic total hip bearing surfaces and their effect on long-term ultra high molecular weight polyethylene wear', *Clin Orthop Rel Res* **294**: 361–78.

Davidson, J. A. and Schwartz, G. (1987) 'Wear, creep, and frictional heat of femoral implant articulating surfaces and the effect on long-term performance – Part I, A review', *Journal of Biomedical Materials Research* **21**: 261–85.

Davidson, J. A., Gir, S. and Paul, J. P. (1988) 'Heat transfer analysis of frictional heat dissipation during articulation of femoral implants', *Journal of Biomedical Materials Research* **22**: 281–309.

Davy, D. T., Kotzar, G. M., Brown, R. H., Heiple, K. G., Goldberg, V. M. et al. (1988) 'Telemetric force measurements across the hip after total arthroplasty', *J Bone Joint Surg Am* **70**: 45–50.

Deirmengian, C. A. and Lonner, J. H. (2008) 'What's new in adult reconstructive knee surgery', *J Bone Joint Surg Am* **90**: 2556–65.

Derbyshire, B., Fisher, J., Dowson, D., Hardaker, C. and Brummitt, K. (1994) 'Comparative study of the wear of uhmwpe with zirconia ceramic and stainless steel femoral heads in artificial hip joints', *Med Eng Phys* **16**: 229–36.

Devine, J. (2006) 'Developing PEEK Polymer as a Bearing Material for Implants', Medical device link. Available from: http://www.devicelink.com/mddi/archive/06/05/007.html (22 June, 2007).

Dowson, D. (1998) *History of Tribology*, London, Bookcraft.

Dowson, D. (2001) 'New joints for the millennium: wear control in total replacements hip joints', *Proc Inst Mech Eng [H]* 335–58.

Edidin, A. A., Pruitt, L., Jewett, C. W., Crane, D. J., Roberts, D. and Kurtz, S. M. (1999) 'Plasticity-induced damage layer is a precursor to wear in radiation-cross-linked UHMWPE acetabular components for total hip replacement', *J of Artroplasty* **14**.

Eftekhar, N. S. and Coventry, M. B. (1992) 'The silver anniversary of The Hip Society. A brief history', *Clin Orthop Relat Res* **285**: 7–10.

Eurotrib.org (2010) 'Tribology'. Available from: http://www.eurotrib.org/tribology (March 2010).

Goldring, S. R., Schiller, A. L., Roelke, M., Rourke, C. M., O'Neil, D. A. and Harris, W. H. (1983) 'The synovial-like membrane at the bone-cement interface in loose total hip replacements and its proposed role in bone lysis', *J Bone Joint Surg Am* **65**: 575–84.

Goldsmith, A. A., Dowson, D., Isaac, G. H. and Lancaster, J. G. (2000) 'A comparative joint simulator study of the wear of metal-on-metal and alternative material combinations in hip replacements', *Proc Inst Mech Eng [H]* **214**: 39–47.

Gomez, P. F. and Morcuende, J. A. (2005) 'Early attempts at hip arthroplasty-1700s to 1950s', *The Iowa Orthopaedic Journal* **25**: 25–9.

Gray, H., Pick, T. P. and Howden, R. (1974) *Gray's Anatomy*, Philadelphia, PA, Running Press Book Publishers.

Heybeli, N. and Mumcu, E. F. (1999) 'Total hip arthroplasty (history and development)', *SDU Tip Fakultesi Dergisi* **6**: 21–7.

Hodge, W. A., Carlson, K. L., Fijan, R. S., Burgess, R. G., Riley, P. O. et al. (1989) 'Contact pressures from an instrumented hip endoprosthesis', *J Bone Joint Surg Am* **71**: 1378–86.

Hsu, S. (1996) 'Ceramic wear maps', *Wear* **200**: 154–75.

Ihn, J. C., Kim, S. J. and Park, I. H. (1993) 'In vitro study of contact area and pressure distribution in the human knee after partial and total meniscectomy', *Int Orthop* **17**: 214–18.

Jacobson, K. (2008) 'Cross-linked Ultra-High Molecular Weight Polyethylene'. Available from: http://www.uhmwpe.unito.it/atti/07%20Jacobson.pdf (Feb 2008).

Kim, K. J., Rubash, H. E., Wilson, S. C., D'Antonio, J. A. and McClain, E. J. (1993) 'A histologic and biochemical comparison of the interface tissues in cementless and cemented hip prostheses', *Clin Orthop Relat Res* **287**: 142–52.

Kurtz, S. M. (2004) *The UHMWPE Handbook*, Elsevier Academic Press.

Lee, K. and Goodman, S. B. (2008) 'Current state and future of joint replacements in the hip and knee', *Expert Rev Med Devices* **5**: 383–93.

Macintosh, D. L. (1958) 'Hemiarthroplasty of the knee using a space occupying prosthesis for painful varus and valgus deformities', *J Bone Joint Surg Am* **40-A**: 1431.

Mancha, H., Gomez, M., Rodriguez, J. L., Escobedo, J., Castro, M. and Mendez, M. (1996) *International Symposium on Reactive Metals: Processing & Applications I: Pure Metals and Alloys*. TMS, Anaheim, California.

Marti, A. (2000) 'Cobalt-base alloys used in bone surgery', *Injury* **31** (Suppl 4): 18–21.

Matijevic, E. (1985) 'Production of monodispersed colloidal particles', *Ann Rev Mater Sci* **15**: 483–516.

McKeever, D. C. (1960) 'Tibial plateau prosthesis', *Clin Orthop* **18**: 86–95.

McKellop, H., Lu, B. and Benya, P. (1992) 'Friction, lubrication and wear of cobalt-chromium, alumina and zirconia hip prostheses compared on a joint simulator', *38th OR*, Washington D.C.

Mcminn, D. J. W. (2003) 'Development of metal/metal hip resurfacing', *Hip International* **13**: S41–53.

Morais, S., Dias, N., Sousa, J. P., Fernandes, M. H. and Carvalho, G. S. (1999) 'In vitro osteoblastic differentiation of human bone marrow cells in the presence of metal ions', *J Biomed Mater Res* **44**: 176–90.

Morais, S., Sousa, J. P., Fernandes, M. H., Carvalho, G. S., De Bruijn, J. D. and van Blitterswijk, C. A. (1998) 'Decreased consumption of Ca and P during in vitro biomineralization and biologically induced deposition of Ni and Cr in presence of stainless steel corrosion, products', *J Biomed Mater Res* **42**: 199–212.

Neu, C. P., Kyriako, K. and Reddi, A. H. (2008) 'The interface of functional biotribology and regenerative medicine in synovial joints', *Tissue Eng Part B Rev* **14**: 235–47.

Older, J. (2002) 'Charnley low-friction arthroplasty', *J Artroplasty* **17**: 675–80.

Piconi, C., Burger, W., Richter, H. G., Cittadini, A., Maccauro, G. et al. (1998) 'Y-TZP ceramics for artificial joint replacements', *Biomaterials* **19**: 1489–94.

Piconi, C. and Maccauro, G. (1999) 'Zirconia as a ceramic biomaterial', *Biomaterials* **20**: 1–25.

Potter, T. A., Weinfeld, M. S. and Thomas, W. H. (1972) 'Arthroplasty of the knee in rheumatoid arthritis and osteoarthritis', *J Bone Joint Surg Am* **54-A**: 1–24.

Rabinowicz, E. (1995) *Friction and Wear of Materials*, Wiley-Interscience.

Ranawat, C. S. (2002) 'History of total knee replacement', *J South Orthop Assoc* **11**: 218–26.

Ranawat, C. S., Flynn, W. F. J. and Saddler, S. (1993) 'Long-term results of the total condylar knee arthroplasty. A 15-year survivorship study', *Clin Orthop*, **286**: 94–102.

Ravaglioli, A. and Krajewski, A. (1992) *Materials for Surgical Use*, London.

Reggiani, M., Tinti, A., Taddei, P., Visentin, M., Stea, S. et al. (2006) 'Phase transformation in explanted highly crystalline UHMWPE acetabular cups and debris after in vivo wear', *J Mol Struct* **785**: 98–105.

Renò, F. and Cannas, M. (2006) 'UHMWPE and vitamin E bioactivity: An emerging perspective', *Biomaterials* **27**: 3039–43.

Renström, P. and Johnson, R. J. (1990) 'Anatomy and biomechanics of the menisci', *Clin Sports Med* **9**: 523–38.

Rieker, C. B. (2003) 'Tribology in total hip arthroplasty – historical development and future trends', *Technology & Service* 1–3.

Rose, R. M., Nusbaum, H., Schneider, H., Ries, M., Paul, I. and Crugnola, A. (1980) 'On the true wear rate of ultra high-molecular-weight polyethylene in the total hip prostheses', *J Bone Joint Surg Am* 537–49.

Saikko, V. (2005) 'A 12-station anatomic hip joint simulator', *Proc Inst Mech Eng [H]* **219**: 437–48.

Saldivar-Garcia, A. J. and Lopez, H. F. (2005) 'Microstructural effects on the wear resistance of wrought and as-cast Co-Cr-Mo-C implant alloys', *J Biomed Mater Res* **74**: 269–74.

Savarino, L., Greco, M., Cenni, E., Cavasinni, L., Rotini, R. et al. (2006) 'Differences in ion release after ceramic-on-ceramic and metal-on-metal total hip replacement', *J Bone Joint Surg Br* 472–6.

Shanbhag, A. S., Jacobs, J. J., Black, J., Galante, J. O. and Glant, T. T. (1995) 'Cellular mediators secreted by interfacial membranes obtained at revision total hip arthroplasty', *J Arthroplasty* **10**: 498–506.

Shen, F. W. and McKellop, H. A. (2002) 'Interaction of oxidation and crosslinking in gamma-irradiated ultrahigh molecular-weight polyethylene', *J Biomed Mater Res* **61**: 430–9.

Tipper, J. L., Ingham, E., Jin, Z. M. and Fisher, J. (2005) 'The science of metal-on-metal articulation', *Current Orthopaedics* **19**: 280–7.

Toni, A., Terzi, S., Sudanese, A., Tabarroni, M., Zappoli, F. A. et al. (1995) 'The use of ceramic in prosthetic hip surgery. The state of the art', *Chir Organi Mov* **80**: 13–25.

Tuan, R. S., Lee, F. Y. T., Konttinen, Y., Wilkinson, J. M. and Smith, R. L. (2008) 'What are the local and systemic biologic reactions and mediators to wear debris, and what host factors determine or modulate the biologic response to wear particles?', *J Am Acad Orthop Surg* **16**: S42–48.

Wang, A., Yue, S., Bobyn, J. D., Chan, F. W. and Medley, J. B. (1999) 'Surface characterization of metal-on-metal hip implants tested in a hip simulator', *Wear* **225–9**: 708–15.

Wikipedia (2007) 'Cross-linked polyehtylene'. Available from: http://en.wikipedia.org/wiki/PEX (25 June 2007).

Wikipedia (2008) 'Fretting wear'. Available from: http://en.wikipedia.org/wiki/Fretting_Wear (10 April 2010).

Wikipedia (2009) 'Erosion'. Available from: http://en.wikipedia.org/wiki/Erosion (1 April 2010).

Wikipedia (2010a) 'Anodizing'. Available from: http://en.wikipedia.org/wiki/Anodizing (7 March 2010).

Wikipedia (2010b) 'Knee replacement'. Available from: http://en.wikipedia.org/wiki/Knee_replacement (26 April 2010).

Wikipedia (2010c) 'Tribology'. Available from: http://en.wikipedia.org/wiki/Tribology (7 March 2010).

Wikipedia (2010d) 'Wear'. Available from: http://en.wikipedia.org/wiki/Wear (7 March 2010).

Willert, H. G., Ludwig, J. and Semlitsch, M. (1974) 'Reaction of bone to methacrylate after hip arthroplasty: a long-term gross, light microscopic, and scanning electron microscopic study', *J Bone Joint Surg Am* **56**: 1368–82.

Willert, H. G. and Semlitsch, M. (1977) 'Reactions of the articular capsule to wear products of artificial joint prostheses', *J Biomed Mater Res* **11**: 157–64.

Williams, D. F. (1986) *Definition in Biomaterials*, ESB, Chester, England.

Willmann, G. (1998) 'Ceramics for total hip replacement – what a surgeon should know', *Orthopedics* **21**: 173–7.

Wimmer, M. A., Sprecher, C., Hauert, R., Täger, G. and Fischer, A. (2003) 'Tribochemical reaction on metal-on-metal hip joint bearings. A comparison between in-vitro and in-vivo results', *Wear* **255**: 1007–14.

Index

Kolibri 360 GmbH
Ferdinand-Jühlke-Straße 7
99095 Erfurt, Deutschland
produktsicherheit@kolibri360.de